THE PARANÁ RIVER BASIN

This book provides insight into the hydrology, ecosystem services and management of water resources in the Paraná River basin, including the importance of water to the socio-economic development of the countries within the watershed.

Running through Brazil, Paraguay and Argentina, the Paraná River and its watershed is home to some of South America's major population centers as well as important ecosystems threatened by development. At the same time, the river is a major resource driving the economies of the nations within its boundaries. This volume examines the impacts of environmental degradation, and the tradeoffs between the energy sector and the maintenance of ecosystem services. In particular, it focuses on the threats from development to sensitive ecosystems within the basin and the challenges of transboundary management of water resources. In addition to presenting wider perspectives on water management, the volume specifically covers water infrastructure, aquatic ecosystems, water quality, geomorphological influences and the impact of climate change. Finally, by assessing each country's current status in meeting the Sustainable Development Goals this volume provides a timely analysis as national governments within the basin are becoming increasingly concerned about the sustainability of the freshwater ecosystems within the Paraná River basin.

This book will be of great interest to students and scholars of water and natural resource management, environmental policy, sustainable development and Latin American studies. It will also be relevant to water management professionals.

Chris D. Metcalfe is a Senior Advisor for Latin America and Caribbean at the UN University Institute for Water, Environment and Health in Canada. He is also a Professor in the School of the Environment at Trent University in Canada.

Mirta L. Menone has been a Lecturer at the University of Mar del Plata (UNMDP) since 2000. Since 2005, she has been a Researcher supported by the National Research Council of Argentina working at the Institute of Marine and Coastal Research at UNMDP.

Pablo Collins is a Researcher supported by the National Research Council of Argentina, working at the National Institute of Limnology, and is also a Professor at the National University of El Litoral, both in Santa Fe, Argentina.

José G. Tundisi has been a Full Professor of Environmental Quality at FEEVALE University in Novo Hamburgo, Rio Grande do Sul State, Brazil since 2009.

EARTHSCAN SERIES ON MAJOR RIVER BASINS OF THE WORLD

Series Editor: Vladimir Smakhtin

Large river basins are dynamic and complex entities. Defined by hydrological boundaries, they are nearly always shared by more than one country. Encompassing a diverse range of landscapes with often huge temporal and spatial variability of resources, they are put to different and often conflicting uses, and managed by a range of institutions and organisations. While an intrinsic part of Nature, many have been extensively engineered and used by people, often with adverse consequences. Each major river basin has its own development trajectory and often fascinating history. Bringing together multidisciplinary teams of experts, this series explores these complex issues, identifies knowledge gaps and examines potential development pathways towards greater sustainability.

The Volta River Basin
Water for Food, Economic Growth and Environment
Edited by Timothy O. Williams, Marloes Mul, Charles Biney and Vladimir Smakhtin

The Ganges River Basin
Status and Challenges in Water, Environment and Livelihoods
Edited by Luna Bharati, Bharat R. Sharma and Vladimir Smakhtin

The Zambezi River Basin
Pathways for sustainable development
Edited by Jonathan Lautze, Xueliang Cai, Everisto Mapedza, Marcus Wishart and Zebediah Phiri

The Aral Sea Basin
Water for Sustainable Development in Central Asia
Edited by Stefanos Xenarios, Dietrich Schmidt-Vogt, Manzoor Qadir, Barbara Janusz-Pawletta and Iskandar Abdullaev

The Paraná River Basin
Managing Water Resources to Sustain Ecosystem Services
Edited by Chris D. Metcalfe, Pablo Collins, Mirta L. Menone and José G. Tundisi

For more information about this series, please visit: https://www.routledge.com/series/ECMRBW

THE PARANÁ RIVER BASIN

Managing Water Resources to Sustain Ecosystem Services

Edited by
Chris D. Metcalfe, Pablo Collins,
Mirta L. Menone, and José G. Tundisi

earthscan from Routledge
Routledge
Taylor & Francis Group
LONDON AND NEW YORK

First published 2020
by Routledge
2 Park Square, Milton Park, Abingdon, Oxon OX14 4RN

and by Routledge
52 Vanderbilt Avenue, New York, NY 10017

Routledge is an imprint of the Taylor & Francis Group, an informa business

British Library Cataloguing-in-Publication Data
A catalogue record for this book is available from the British Library

Library of Congress Cataloging-in-Publication Data
Names: Metcalfe, Christopher D., 1951- editor.
Title: The Paraná River Basin : managing water resources to sustain ecosystem services / edited by Chris D. Metcalfe, Pablo Collins, Mirta L. Menone, José G. Tundisi.
Description: Abingdon, Oxon ; New York, NY : Routledge, 2020. | Series: Earthscan series on major river basins of the world | Includes bibliographical references and index.
Identifiers: LCCN 2020002643 (print) | LCCN 2020002644 (ebook) | ISBN 9780367277048 (hardback) | ISBN 9780429317729 (ebook)
Subjects: LCSH: Water resources development—Paraná River Watershed (Brazil-Argentina) | Water supply—Paraná River Watershed (Brazil-Argentina) | Sustainable development—Paraná River Watershed (Brazil-Argentina) | Paraná River Watershed (Brazil-Argentina)—Environmental conditions.
Classification: LCC HD1696.P37 P37 2020 (print) | LCC HD1696.P37 (ebook) | DDC 333.9100982/2—dc23
LC record available at https://lccn.loc.gov/2020002643
LC ebook record available at https://lccn.loc.gov/2020002644

ISBN: 978-0-367-27704-8 (hbk)
ISBN: 978-0-429-31772-9 (ebk)

Typeset in Bembo
by Swales & Willis, Exeter, Devon, UK

CONTENTS

CHAPTER PHOTOS

1 The Paraná River in the Upper basin.
Photograph by Rodrigo Soldon obtained through an agreement with Flickr.com

2 The Paraná River on the border between Brazil, Argentina and Paraguay.
Photograph provided by Antonio Cezar Leal of São Paulo State University, Presidente Prudente campus

3 The Paraná River from the Friendship Bridge (Ponte de Amizade, Puente de la Amistad) linking the city of Foz de Iguaçu in Brazil with the city of Ciudad del Este in Paraguay.
Photograph obtained through an agreement with Flickr.com

4 The Puerto Primavera dam located on the Upper Paraná River.
Photograph provided through an agreement with creativecommons.org

5 The Pantanal wetland.
Photograph by Uwe Bergwitz obtained through an agreement with Shutterstock.com

6 The Paraná River and Iguazu Falls.
Photograph provided by Ricardo Zig Koch Cavalcanti of the Brazilian Water Resources Agency

7 The Paraná River showing agricultural development.
The photograph was taken by Alexandre Marchetti of Itaipú Binacional and it was provided by Dieison André Moi by authorization from Itaipú Binacional

8 The suspension bridge across the Paraná River between São Paulo and Mato Grosso do Sul states in Brazil.
Photograph obtained through an agreement with Shutterstock.com

9 The source of drinking water in a rural community in eastern Bolivia.
Photograph taken by Ruslana Iurchenko and obtained through an agreement with Shutterstock.com. Note that there is a signed model release on file with Shutterstock Inc

NOTES ON CONTRIBUTORS

Renata Ribeiro de Araújo holds a Chemical Engineering degree (1994), a Master's degree in Chemical Engineering (1997) and a PhD in Ecology of Continental Aquatic Environments (2003), all from the State University of Maringá in Brazil. She is currently a PhD Assistant Professor with Júlio de Mesquita Filho, São Paulo State University (UNESP) at the Presidente Prudente campus in Brazil. She has experience in the environmental area in limnology, ecosystem ecology, environmental education in water resources, environmental assessment and recovery of degraded areas. She serves as coordinator of the Graduate Program in Geography, Professional Master's Degree in Water Resources and Environment of FCT/UNESP and participates on the São Paulo State Drainage Basin Committee. She has published 26 articles and 10 book chapters and has edited 3 books in her field of expertise. She has supervised students for 10 Master's degrees, 9 specializations, 64 scientific initiations, and 31 undergraduate theses.

João Paulo Lyra Fialho Brêda holds an undergraduate degree in Environmental Engineering from the Federal University of Alagoas and a Master's degree in water resources and environmental sanitation from the Institute of Hydraulic Research (IPH) at the Federal University of Rio Grande do Sul (UFRGS) in Brazil. Currently, he is a PhD candidate at this same institute. His main research topics are continental-scale hydrological modelling, climate change impacts on water resources in South America and hydrological data assimilation methods applied with remote sensing applications.

José Esteban Castro is an Emeritus Professor of Sociology at Newcastle University in the United Kingdom. He is a Principal Researcher at the National Scientific and Technical Research Council (CONICET) in Argentina, linked to the National University of General Sarmiento (UNGS) in Argentina, and a regular visiting scholar at universities in Brazil, Colombia, Mexico, Spain, and the United Kingdom. He has done extensive research on structural inequality and injustice connected with the politics of water in Europe and Latin America, and leads the international WATERLAT-GOBACIT Network, which is dedicated to teaching, research and practical action on water politics and management. Dr. Castro is a Fellow of the Norbert Elias Foundation and a Corresponding Member of the Mexican Academy of Sciences.

Pablo Collins is a Researcher supported by the National Research Council of Argentina (CONICET), working at the National Institute of Limnology (INALI) and he is also a Professor at the National University of El Litoral (UNL), both in Santa Fe, Argentina (i.e. INALI-CONICET-UNL). Dr. Collins specializes in the study of limnology and environmental contamination. He has developed an international reputation for his research on freshwater macro-crustaceans, evaluating their biology and ecology. Moreover, his research includes studies of the toxic effects of contaminants in this zoological group. He has published over 116 articles in the peer-reviewed literature on these research topics. Dr. Collins also has experience in research and capacity building in integrated watershed management due to his INALI affiliation.

Walter Collischonn holds an undergraduate degree in Mechanical Engineering, and a Master's and PhD degrees in water resources and environmental sanitation from the Federal University of Rio Grande do Sul (UFRGS), in Porto Alegre, Brazil. Currently, Dr. Collischonn is an Associate Professor at the Institute of Hydraulic Research (IPH) at the same university and his main field of research is hydrology. He developed the 'MGB' large-scale hydrological model and he has conducted research on large-scale hydrology of South American river basins, including streamflow and flood forecasting and the impacts of climate change on water resources. He has supervised 30 Master's and PhD candidates and he has published over 50 articles in peer-reviewed international journals.

Antonio Cezar Leal is a Geographer, with a Master's degree in Geosciences and Environment from the São Paulo State University Brazil (1996), and a PhD in Geosciences from the University of Campinas, Brazil (2000). Since 1996, he has served as a Professor at the School of Science and Technology of São Paulo State University (UNESP) at the Presidente Prudente campus in Geography, Water Resources and Environment and in the management and regulation of Water Resources. He has collaborated on research activities alongside researchers from the University of Havana in Cuba, University of Lisbon in Portugal, and Valley University in Colombia. He coordinated an undergraduate and postgraduate program, institutional projects in collaboration with other universities and was an Advisor to the Provost's Office for University Extension at UNESP. He participates in Committees of the São Paulo State Drainage Basins for the River Basin Committee (CBH) of Paranapanama (Interstate), and the State Council of Water Resources of São Paulo. He is the Coordinator of the Technical Chamber (CT) for Management Instruments of CBH Paranapanema, and Deputy Coordinator of the CT for Ecological Restoration for the CBH, Paranapanema Point (i.e. Pontal do Paranapanema). He is a holder of a Research Productivity Scholarship awarded by the National Research Council (CNPq) of Brazil, Level 2. Dr. Leal has published 71 articles, 11 books (9 edited), and 36 book chapters. He has supervised 25 Master's students and 22 doctoral students, 56 scientific initiations, and 122 undergraduate theses. He develops research and extension activities in water resource planning and management in drainage basins and in urban solid waste management, focusing on waste sorting and organization of recyclable waste pickers, with the support of public and private institutional partnerships and the regional community.

Mercedes R. Marchese is a Researcher with the National Research Council of Argentina (CONICET), working at the National Institute of Limnology (INALI), affiliated with the National University of El Litoral (UNL) in Santa Fe, Argentina. She was a Professor Titular at the Faculty of Humanities and Sciences (FHUC) at UNL. Her current research includes the ecology of benthic invertebrates in large floodplain rivers, the study of food webs using stable isotopes and benthic meta-community structures in the Paraná River floodplain. Dr. Marchese also has experience studying the effect of environmental perturbations on the function of invertebrate assemblages, and the use of *in situ* and laboratory bioassays to assess toxicity and bioaccumulation. She also has expertise in the taxonomy of freshwater oligochaetes from South America and their ecology, in addition to their relationships to different aquatic habitats from the Paraná River system. She has published over 77 articles in peer-reviewed journals.

Cláudio Antonio Di Mauro received a Bachelors and a Licentiate degree in Geography from the Auxilium School of Philosophy, Sciences and Letters in Lins, Brazil (1971) and a Master's degree (1981) and a PhD (1989) in Physical Geography from the School of Human Sciences of the University of São Paulo (USP) in Brazil. He serves as an Assistant in the Geomorphology Division for the RADAM Brasil remote sensing project. He entered the Master's degree program at USP under the supervision of Prof. Dr. Aziz Ab'Sáber and defended his Master's degree and Doctoral thesis under the supervision of Prof. Dr. Olga Cruz. He acted and retired as a Professor in the Júlio de Mesquita Filho with São Paulo State University (UNESP) at the Rio Claro campus in Brazil, where he coordinated the Geography program for two years. In 1996, he was elected Mayor of the Municipality of Rio Claro, Brazil. He was re-elected in 2000, serving his term until the end of 2004. As Mayor of Rio Claro, he participated in the first edition of the World Social Forum in Porto Alegre, Brazil and the Local Authorities Forum. He headed for six years the Piracicaba, Capivari and Jundiaí River Basin Committee for São Paulo State, and later at the federal level. He worked as an Advisor at the National Water Agency (ANA) and as a consultant to UNESCO. In 2008, he was accepted by the Federal University of Uberlândia (UFU) as an Associate Professor in the Institute of Geography (IG). He is an active member of the Paranaíba River Basin Committee, representing Technical and Research Institutions on behalf of UFU. He is accredited in the Professional Graduate Program in Geography at UNESP, Presidente Prudente campus, and in the Prof-Water Program, coordinated by UNESP Ilha Solteira.

Hamid Mehmood is a Senior Researcher in Hydro-informatics and Information Technology at the United Nations University Institute for Water Environment and Health (UNU-INWEH) based in Hamilton, Ontario, Canada. He holds Master and PhD degrees in Remote Sensing and GIS from the Asian Institute of Technology (AIT) in Bangkok. Mehmood worked in research and technical programs on hydro-informatics and information technology for over 20 years. He has worked on technology transfer projects in more than 15 developing countries and was involved in the implementation of the Hyogo Framework for Action of the UN Economic Commission for Asia and Pacific (UNESCAP). He also served as an Assistant Professor at leading universities in Pakistan. His areas of interest and expertise include application of hydro-meteorological sensors as Internet of Things, deploying blockchain in developing countries as a primary sensor data sharing and storing mechanism, solving hydro-meteorological data gaps using new information technologies, such as cloud computing for earth observations, and the use of open source online learning platforms as part of a hybrid capacity building approach.

Mirta L. Menone has been a Lecturer at the University of Mar del Plata (UNMDP) since 2000. Since 2005, she has been a Researcher supported by the National Research Council of Argentina (CONICET) working at the Institute of Marine and Coastal Research (IIMYC) at UNMDP. She completed her PhD studies at UNMDP and took a Post-doctoral position at the

IGB Leibniz Institute of Freshwater Ecology and Inland Fisheries in Berlin (Germany) through support from a Humboldt Research Fellowship (AvH Foundation). She has supervised more than 20 undergraduate fellowship students and 6 PhD students. She is currently working on biomonitoring of current-use pesticides in aquatic ecosystems and the use of early warning systems (biochemical and genetic biomarkers) for biomonitoring. She has published 35 scientific papers in peer-reviewed journals.

Chris D. Metcalfe is a Senior Advisor with the United Nations University Institute for Water, Environment and Health (UNU-INWEH) located in Hamilton, Ontario, Canada. He is also a Professor in The School of the Environment and the Director of the Institute for Watershed Science at Trent University in Peterborough, Ontario, Canada. Dr. Metcalfe has developed an international reputation for his research on the fate and toxic effects of contaminants in the environment, and he has published over 250 articles in the peer-reviewed literature on this research topic. Dr. Metcalfe also has experience in research and capacity building in integrated watershed management, with a focus on community-based projects with Indigenous peoples in Canada, and in countries within the region of the Caribbean and Latin America.

Oscar Orfeo received his PhD in Geology from the National University of La Plata in Argentina. He is a Researcher supported by the National Council of Scientific and Technical Research (CONICET) as Director of the Centre of Applied Littoral Ecology (CECOAL), and he is a Professor Category 1 at the National Northeastern University (UNNE) in Corrientes City, Argentina. He has published numerous articles in international scientific journals related to the hydro-sedimentology and drainage patterns of fluvial systems in subtropical climates. He has given more than 50 guest lectures inside and outside Argentina, most of them related to the Paraná River basin.

Rodrigo Cauduro Dias de Paiva holds an undergraduate degree in Civil Engineering from the Federal University of Santa Maria, Brazil, and Master's and PhD degrees in water resources and environmental sanitation from the Institute of Hydraulic Research (IPH) at the Federal University of Rio Grande do Sul (UFRGS), where he is currently an Associate Professor. Dr. Paiva conducts research on hydrology and remote sensing. He developed a large-scale hydrologic hydrodynamic model for the Amazon basin and conducted research on the predicted impacts of climate change on water resources and on the changes in river regimes caused by future construction of dams in the Amazon. He has advised eight Master's and PhD students and he has published over 30 articles in peer reviewed journals.

Daniel Parsons is a Professor at the University of Hull, UK, and he is an active researcher in areas related to fluvial, estuarine, coastal and deep marine sedimentary environments, exploring responses of these systems to climate and environmental change. He has research interests in anthropogenic disturbances to these systems and determining necessary societal adaptations to mitigate the impact of change. He has published numerous papers in scientific journals devoted to fluvial geomorphology.

Alejandro Pastori is a Professor in Public International Law at the School of Law of the University of the Republic in Montevideo, Uruguay. Dr. Pastori's main field of research has been in Regional Integration within South America. In recent years, he has studied the legal framework of the River Plate Basin and he recently published a book on this topic. Besides his academic career, he has been a Senior Legal Advisor for the Latin American Integration Association (ALADI) and a consultant for several international organizations on topics within his expertise.

Wagner Costa Ribeiro holds an undergraduate degree in Geography from the University of São Paulo (1986) and a PhD in Human Geography from the University of São Paulo (1999) in Brazil. He is a Full Professor in the Department of Geography, Graduate Program in Human Geography and Graduate Program in Environmental Science at the University of São Paulo (USP), which he headed from 2006 to 2008. He was Coordinator of the Research Group in Environmental Sciences at the Institute of Advanced Studies of USP from 2008 to 2012. He obtained his Free Teaching Certificate in 2004, also at USP. He completed his postdoctoral studies at the University of Barcelona in Spain and he has served as Visiting Professor at the University of Salamanca, University of Seville, and University of Caldas in Colombia. He is a coordinator at the Cross-Border Water Thematic Area of the WATERLAT-GOBACIT research network. He is a Research Productivity Scholarship Holder at the National Research Council of Brazil (CNPq), Level 1C. He has experience in the field of Geography, with an emphasis on Political Geography and the environment. His main research topics are environmental public policies, international relations and the environment, water resource management, and international environmental order. Dr. Ribeiro is the author of several academic articles, the books *A ordem ambiental internacional* (2001 and 2005 – second edition) and *Geografia política da água* (2008), and the didactic collection, *Por dentro da Geografia* and *Geo Conecte Live*.

Paulo Cesar Rocha holds a PhD in the Ecology of Continental Aquatic Environments from the State University of Maringá (2002) in Brazil. He held a Post-doctoral position in the Ecology Program for Continental Aquatic Environments in the School of Environment at the University of Liverpool (2015). He is currently an Assistant Professor at the School of Science and Technology of the São Paulo State University (FCT/UNESP) in Brazil, and a Research Productivity Assistant at the National Research Council of Brazil (CNPq), Level 2. He has experience in geosciences/physical geography, with an emphasis on geomorphology and fluvial dynamics, hydrological regimes in hydrographic basins, morphology and dynamics of fluvial plains, limnology, and fluvial sedimentology. He is the Coordinator of the Laboratory of Geology, Geomorphology and Water Resources of FCT/UNESP. He is also the Coordinator of the Graduate Program in Geography for a Professional Master's Degree in Water Resources and Environment at FCT/UNESP. His international experience is linked to internships and student exchange programs at the University of Liverpool, UK and student exchange programs with the University of Wollongong, Australia. Dr. Rocha has published 48 articles and 32 book chapters in his field of expertise. He has supervised 11 Master's dissertations, 7 doctoral theses, and 2 postdoctoral fellows. His current research is aimed at studying hydrodynamic connectivity between river basin elements and river channels in an approach to evaluate the impacts of sugarcane farming expansion on river basins and river channels.

Francesco Sindico is Co-Director and Founder of the Strathclyde Centre for Environmental Law and Governance at the University of Strathclyde Law School in Glasgow, Scotland, UK. Dr. Sindico has over ten years of research and policy experience in the field of international water law, with a particular focus on transboundary aquifer law and policy. His work has focused on the management of the Guarani Aquifer System and he has published several papers on the Guarani Aquifer Agreement.

Vinícius Alencar Siqueira holds an undergraduate degree in Environmental Engineering and a Master's degree in Water Resources and Environmental Sanitation, both from the Institute of Hydraulic Research (IPH) at the Federal University of Rio Grande do Sul (UFRGS). Currently, he is a PhD candidate at the same university. His main research topics are continental-scale hydrological modelling, streamflow and flood forecasting in large basins of South America, and geoprocessing and remote sensing applied to water resources. He has published ten articles in peer-reviewed journals.

José Stevaux received his PhD in Geology from the University of São Paulo (USP) in Brazil. He is a Visiting Professor at the Federal University of Mato Grosso do Sul (UFMS) at the Campus of Três Lagoas in Brazil. He is a Researcher A1 at the National Research Council of Brazil (CNPq), and co-leader of the Tropical Rivers Working Group of the Association of Geomorphologists and the Tropical Rivers, IGCP 582 project of UNESCO. He has published 107 papers in scientific journals, as well as a book on fluvial geomorphology.

Franco Teixeira de Mello is a Professor from the Eastern Regional University Centre at the University of the Republic (UdelaR) in Uruguay. Since 2013, he has been a researcher in the postgraduate Program for Development of Basic Sciences (PEDECIBA) in Geosciences at UdelaR, and since 2015, he has been a researcher in the postgraduate program of PEDECIBA. His international experience includes cooperation in projects with researchers from Argentina, Brazil, Colombia, Venezuela, Ecuador, the United States, Denmark, Portugal, Spain and China. He also conducts Antarctic research. He has supervised nine Masters-level students and two Doctoral-level students. His current research includes freshwater ecology, biomonitoring of fish and invertebrates, the dynamic of pesticides, and organic decomposition in inland aquatic ecosystems, as well as studies of micro- and macro-plastic contamination. Dr. Teixeira de Mello has published 50 scientific papers in peer- reviewed journals and is an Associate Editor for three journals.

José Galizia Tundisi was a Full Professor of Environmental Sciences in the School of Engineering at the University of São Paulo in São Carlos, São Paulo State, Brazil since 1996 (now retired). Since 2009, Dr. Tundisi has been a Full Professor of Environmental Quality at FEEVALE University in Novo Hamburgo, Rio Grande do Sul State, Brazil. His main research activities are the ecology and limnology of reservoirs. His current research includes studies of the impacts of cold fronts on reservoirs in the Southeast region of Brazil and the response of the biological communities to these weather patterns. He has supervised 42 MSc thesis students and 32 PhD thesis students, and he has published 450 papers in peer-reviewed scientific journals and 37 books. He is President of the International Institute of Ecology, a non-profit organization situated in São Carlos, Brazil. He is a member of the Brazilian Academy of Sciences and was President of the National Research Council of Brazil (CNPq) over the period from 1995 to 1999. Since 1991, he has been a staff member of the Ecology Institute with Excellence in Ecology at Oldendorf in Germany. He is the focal point for Brazil in the Water Committee of the Inter-American Network of Academies of Sciences.

Takako Matsumura-Tundisi has been a Full Professor of Ecology at the Federal University of São Carlos in São Paulo State, Brazil since 1992 (now retired). Since 1998, she has been the Scientific Director of the International Institute of Ecology, a non-profit organization located in São Carlos, Brazil. Her area of specialization is aquatic biodiversity and limnology of reservoirs and natural lakes. She has developed several research projects on the reservoirs of São Paulo State and other Brazilian States, which have been funded by the National Research Council of Brazil (CNPq), the Research Foundation of São Paulo State (FAPESP) and the Project Financing Company of the Government of Brazil (FINEP). Her current research includes studies of the geographical distribution of zooplankton in reservoirs and their relationships with eutrophication, pollution and water quality. She has supervised 25 MSc thesis students, 25 PhD thesis students and 150 undergraduate students for scientific initiation and specialization. She has published 125 papers in peer-reviewed journals and 10 books. She is Chief Editor of *Brazilian Journal of Biology* and a member of the São Paulo State Academy of Sciences.

FOREWORD

The Paraná River Basin extends over large areas of Brazil, Paraguay and Argentina, including the basin of the Paraguay River. The Paraná River Basin has a large socio-economic influence due to the fact that it is located in one of the most developed, industrialized and densely populated regions of South America.

Besides the large population and many economic activities within the basin (industry, navigation, agriculture, livestock, fish farming, mining, recreation and tourism), the huge demand for hydro-power generation increases the dependence of countries on the basin's water resources. An example of this is the Itaipu Binational Power Plant, one of the biggest hydroeletric plants in the world. Located along the upper reach of the Paraná River along the Brazil-Paraguay border, it is one of the most important sources of electricity consumed by both countries.

Because of this scenario of intricate socio-economic interdependence among the basin's water resources and its multiple users, this book makes a key contribution to deepen our reflections about the future of this extremely important South American river basin. In describing the social, political and institutional aspects, the book points out the existing institutional weaknesses and offers clear pathways to improve the instruments of governance to achieve a more balanced use of water resources.

As a transboundary river basin shared by Brazil, Paraguay and Argentina, the establishment of a consistent and long-term cooperative relationship among the key stakeholders is essential. As the book rightly points out, the combined effort required to develop an integrated and efficient water resources management framework among federal entities and countries is the only choice to achieve an equitable and fair sharing of the benefits that arise from the water resources. This is an extremely important global issue since nearly 56% of the world's land mass lies in river basins that are shared by two or more countries.

The challenges and opportunities for sustainability of water resources are also appropriately addressed in this book by describing issues related to the basin's hydrological cycle, water quantity/quality, biodiversity, geomorphological richness, surface and groundwater resources, water infrastructure and climate changes impacts.

Regarding pollution in the basin from diffuse sources and from domestic/industrial effluents, solutions go beyond the issue of sanitation in the basin, reflecting the situation regarding sewage collection and treatment in these countries, in line with Sustainable Development

Goal 6: (Clean Water and Sanitation) of the UN 2030 Agenda for Sustainable Development. Relevant improvements in this matter would be strengthened with a greater involvement of policymakers and effective use of existing regulatory and institutional instruments.

The opinions of the experts who are authors in the book about these urgent matters comprise a broad scientific diagnosis for those who are in search of a deep understanding about the water resources, environment and multiple stakeholders in the basin. The wide range of knowledge included in the book provides an objective and comprehensive synthesis of much of the South American experience, and also frames this book as a resource for addressing the complex challenges faced in many river basins worldwide.

Benedito Braga CEO of SABESP (Water utility of the State of São Paulo, Brazil); Honorary President of the World Water Council; Former President of the International Water Resources Association

ACRONYMS AND ABBREVIATIONS[1]

ADT	Cross-Border Diagnostic Analysis
ALADI	Latin American Integration Association
ANA	National Water Agency, Brazil
ANDE	National Electrical Administration, Brazil
AWR	Arsenic in Water Research Group, Argentina
cBOD	Carbonaceous Biological Oxygen Demand
CARP	Administrative Commission for the Rio de la Plata
CARU	Administrative Commission for the Uruguay River
CBH	River Basin Committee
CEMADEN	National Monitoring and Alerting Centre for Natural Disasters, Brazil
CIC	Intergovernmental Coordinating Committee of the Countries of the La Plata Basin
CIH	Intergovernmental Waterway Committee
CMIP5	Coupled Modesl Intercomparison Project 5
CNPq	National Research Council, Brazil
CNRH	National Council of Water Resources, Brazil
COBINABE	Binational Commission for Development of the High Basin of the Bermejo River
COHIFE	Federal Water Council, Argentina
COMIP	Argentine-Paraguayan Mixed Commission of the Paraná River
CONAM	National Environment Council, Paraguay
COREBE	Regional Commission of the Bermejo River
CURI	Uruguayan Council for International Relations, Uruguay
DAEE	Department of Water and Electricity for the State of São Paulo, Brazil
DSD	Department of Sustainable Development, Organization of American States
EBY	Yaciretá Binational Entity
ECLAC	The Economic Commission for Latin America and the Caribbean
EIA	Environmental Impact Assessment
Eletrobrás	Brazilian Electrical Centre.
FAPESP	Research Foundation of São Paulo State, Brazil

1 Note that Spanish and Portuguese acronyms are translated into English.

FINEP	Project Financing Company, Government of Brazil
FONPLATA	Financial Fund for the Development of the Rio de la Plata Basin
GAA	Guarani Aquifer Agreement
GBD	Global Burden of Disease
GCM	Global Climate Model
GEMI	The UN-Water Organization for Integrated Monitoring of Water and Sanitation
GEMS/Water	The Global Environment Monitoring System for Freshwater
GEMStat	The water quality data base operated under the auspices of GEMS/Water
GHG	Greenhouse Gas
GLAAS	The UN-Water Organization for Global Analysis and Assessment of Sanitation and Drinking Water
HR	Hydrographic Region
HRU	Hydrological Response Units
IANAS	Inter-American Network of Academies of Sciences
IIEGA	International Institute of Ecology and Environmental Management
IMI	UN-Water Integrated Monitoring Initiative for SDG 6
INA	National Water Institute, Argentina
IPCC	Intergovernmental Panel on Climate Change
ITAIPU	Itaipu Binational Entity
IWRM	Integrated Water Resources Management
JMP	WHO/UNICEF Joint Monitoring Programme for Water Supply, Sanitation and Hygiene
LPB	La Plata Basin
MGB	Large-scale Hydrological Model
MMA	Ministry of the Environment, Brazil
MMAyA	Ministry of Environment and Water, Bolivia
NGO	Non-Governmental Organization
OAS	Organization of American States
PAE	Strategic Action Program for the La Plata Basin
PRB	Paraná River Basin
RCP	Representative Concentration Pathways
SABESP	Basic Sanitation Company of the State of São Paulo, Brazil
SDO	Sustainable Development Objectives
SEAM	Secretary of Environment of the National Government, Paraguay
SGD	Sustainable Development Goals
SINGREH	National Water Resources Management System, Brazil
TCP	Treaty of the La Plata Basin
UFSCar	Federal University of São Carlos, Brazil
UGRH	Water Resource Management Unit
UHE	Hydroelectric Power Plant
UNEP	United Nations Environmental Programme
UNSD	United Nations Statistics Division
UPRB	Upper Paraná River Basin
WRAU	Water Resources Analysis Unit, Brazil
WRI	World Resources Institute
WWAP	World Water Assessment Program

1

The Paraná River Basin and its ecosystem services

Chris D. Metcalfe and Mirta L. Menone

Key messages

- The Paraná River Basin in South America is the second largest on the continent after the Amazon, and the 13th largest river system in the world, with a basin area of approximately 2.6×10^6 km^2 and a population in the basin of approximately 100 million people.

- The Paraná River and its tributaries and wetlands are of great economic, ecological and cultural importance to the region and to the world.
- Because of rapid urbanization, expanding agriculture and the development of hydroelectric projects, this hydrologic system is under stress.
- These changes within the basin have impacted ecosystem services, including habitat, cultural, regulating and provision services. Strategies are needed to maintain these ecosystem services in order sustain the growing human populations within the region.
- One of the impacts on ecosystem services has been loss of habitat, which has reduced biodiversity within the basin. Climate change and invasive species have also negatively impacted the basin.
- Concurrent with the loss of biodiversity has been an increase in the numbers of endangered species within the basin. These species include the critically endangered fish species, *Brycon orbignyanus* (known locally as *Piracanjuba*), the endangered Four-eyed frog (*Pleurodema kriegi*) and the Neotropical river otter (*Lontra longicaudis*), which is 'data deficient' in conservation status.
- Invasions of exotic species, such as the Asian clam (*Corbicula fluminea*) into the Paraná basin may also alter the ecology, including negative effects on ecosystem services.
- In this book, experts in a range of disciplines review the hydrology, geomorphology, ecology, water quality, water resource management, international cooperation and prospects for meeting the Sustainable Development Goal for water and sanitation in the basin.

Introduction

The Paraná River and its tributaries are part of a massive watershed that covers much of the south-central part of South America. The drainage basin of the Paraná River is the second largest on the continent, second only to the Amazon River, and is the 13th largest river system in the world. The Paraná River delta is a huge forested marshland located northwest of the city of Buenos Aires in Argentina. The upper part of the basin includes the ecologically important Cerrado, Pantanal and Atlantic Forest regions of Brazil. The wetlands that extend from the Pantanal and upper Paraguay rivers to the alluvial valley of the mid and lower parts of the Paraná River are the largest freshwater wetland system on the planet.

The Lower Paraná is navigable and is used as a waterway linking inland cities to the Atlantic Ocean. Large oceangoing vessels can travel up the Paraná River as far as the cities of Santa Fe or Paraná during the wet season, but beyond that point only small riverboats can reach into Brazil. In the Upper Paraná, many dams have been built for power generation. The Itaipu Dam in Brazil is the second largest hydroelectric installation in the world, next to the Three Gorges Dam in China, and its reservoir has flooded approximately 100,000 ha of land (Stevaux et al., 2009). Some sections of the Upper Paraná can be used for boat traffic because locks were constructed that allow navigation by shallow draft vessels around the hydroelectric dams, but other parts of the upper basin are not navigable. Much of the length of the Paraná and its tributaries has been impacted negatively by urbanization, industry and agricultural development.

Main physiographic features of the basin

The basin originates in the southern part of Brazil and flows in a southerly direction through parts of Paraguay and Argentina, with some contribution from headwaters located in the Andean region of eastern Bolivia (Figure 1.1). The area of the Paraná River Basin is approxi-

Figure 1.1 The Paraná River Basin and major urban centers within the watershed. The southern and eastern boundaries of the contiguous La Plata basin are also shown. Hamid Mehmood compiled the geospatial data in ArcMap format and Erin Hayward prepared the map using those data.

mately 2.6×10^6 km^2 (Iriondo, 2007), and it partly serves as the international boundary between Brazil and Paraguay and the border between Argentina and Paraguay. The largest subregion of the Paraná-Paraguay basin is the Gran Chaco, a low interior plain with a dry climate that was formed by sediment from the Andes. In the northern part of the basin, at the junction of Paraguay, Bolivia and Brazil, lies the low and swampy Pantanal wetland. The eastern basin includes the Mato Grosso tropical grasslands. The Paraná Plateau in the southern area of the basin, primarily in the eastern part of Paraguay, has rich volcanic soil. Tropical and subtropical climates dominate in most of the basin.

The sub-basin that drains into Paraguay River in the northwestern part of the basin is often considered a separate part of the system, but in this book, the Paraguay River and its tributaries are considered part of the Paraná River Basin (Figure 1.1). The part of the system that comprises the Paraná River is subdivided into three sections that differ in their hydrologic and ecological characteristics (Isupova and Mikhailov, 2018; Iriondo, 2007; Orfeo and Stevaux, 2002):

- The Upper Reach, which is about 3,480 km in length, extends from the source of the Paranaiba River to the confluence of the Paraguay River with the Paraná River, which is a short distance upstream of the city of Corrientes. This part of the system is divided into several sub-basins contributing to multiple tributaries.
- The Middle Reach, which is 820 km in length and extends from the confluence of the Paraguay River and Paraná River to the inflow of the Salado River near the cities of Santa Fe and Paraná.
- The Lower Reach, which is about 500 km in length and extends from the mouth of the Salado River to the point of discharge of the Paraná River into the Rio de la Plata estuary. This section also includes the Delta, which begins at Rosario city at the furthest upstream influence of the Atlantic tides. The Delta extends along its main (northern) branch for about 222 km and along the southern branch for about 207 km (Isupova and Mikhailov, 2018).

The Delta discharges into the Río de la Plata, which is a shallow coastal plain estuary on the east coast of South America. The estuary receives freshwater from both the Paraná River Basin and the Uruguay River (Figure 1.1). The hydrological cycles in the Paraná River and the Uruguay River affect the extent of saltwater intrusion into the estuary and the sediment loads entering the estuary. These are important influences on the major cities and shipping ports of Buenos Aires in Argentina and Montevideo in Uruguay. For instance, frequent dredging is required to keep the port of Buenos Aires open to ocean going vessels. Navigation upstream into the Paraná River is made hazardous by shifting shoals produced by silt deposition. The balance between the influences of saltwater and freshwater is also important to the maintenance of coastal wetlands, such as the ecologically important Samborombón wetland on the south coast of the Rio de la Plata estuary (Carol et al., 2013). Figure 1.1 also shows the southern and eastern margins of the Rio de la Plata basin, since several international agreements related to water management in the region include this basin along with the Paraná River Basin.

The hydrological cycle of the Paraná River Basin includes a flood period in the first six months of the year, with maximum discharge peaks in March to April, while minimum flows occur in September to October. The Upper Reach of the Paraná, with an average discharge of approximately 12,000 m^3 sec^{-1} governs the hydrologic regime characteristics of the middle and lower reaches (Drago and Amsler, 1998). According to Isupova and Mikhailov (2018), the average water discharge decreases along the Middle Paraná River due to losses to evaporation and consumptive use. The morphology of the Paraná River Basin is characterized by extensive lotic (flowing water) and lentic (still water) ecosystems that are eroded, filled or enlarged in each flood event. The fluctuations in water levels within these water bodies (i.e. 'flood pulsing') are linked ultimately to the dynamics and productivity of the main river channel (Vega et al., 2014; Ashworth and Lewin, 2012).

The importance of the Paraná River Basin to the environment lies with the high biodiversity of fluvial fauna and alluvial vegetation within the tropical, sub-tropical and more temperate regions of the watershed (Junk et al., 2014; Barletta et al., 2010; Quirós et al., 2007; Junk et al., 2006). The basin also plays a key role in both the storage of sediment at a continental scale and the transport of sediments from the continent to the Atlantic Ocean at a global scale (Latrubesse and Abad, 2018).

Basin development and water management challenges

The Tupi-Guarani Indigenous peoples predominantly inhabited the basin before Europeans arrived. Minor development by European colonists occurred in eastern Paraguay and the Misiones province of northern Argentina in the 17th and 18th centuries. By the early 19th century, the basin's economy relied on subsistence agriculture but also included some commercial livestock ranches in the Entre Ríos province of Argentina. In the 20th century, the basin began to be settled intensively. Cattle grazing became the land's primary economic use, but intensive irrigation was introduced into the Gran Chaco region for coffee cultivation. Lumbering also became a major industry in the large stand of commercial-grade pines on the Paraná Plateau. However, poor transportation in the region continued to discourage development and hindered exploitation of natural resources.

In the 20th century, international conflicts erupted over the hydroelectric potential of the Paraná and Paraguay rivers. Many regard the 15-month border conflict that erupted between Brazil and Paraguay between 1965 and 1966 over access to the hydroelectric potential of the basin as a key geopolitical event that was a springboard for Brazil's rising power and the economic marginalization of Paraguay and Argentina (Blanc, 2018). In 1973, Brazil and Paraguay signed the hydroelectric agreement at Itaipú, and hydroelectricity has since been the essential economic focus for development within the Upper Basin. With the availability of cheap hydroelectric power, there has been rapid industrialization and urbanization within this part of the basin.

The primary stressors within the Paraná River Basin are a result of urbanization and the expansion of agriculture and hydroelectric development. Of the countries that are situated in the Paraná River Basin, Brazil is the largest. The country's population has grown rapidly since the 1950s, with a four-fold increase to its present population size of 209 million. With a Gross Domestic Product of approximately $2 trillion, Brazil's economy is ranked eighth in the world, although a significant proportion of the people still live in poverty (Nakabashi et al., 2013). Argentina, with a population of approximately 44 million, is considered an upper-middle income country with a Gross Domestic Product of $600 billion. Paraguay is a landlocked country with a population of 6.7 million and a Gross Domestic Product of $26 billion. Approximately 30% of the population of Paraguay lives below the national poverty line.

As a result of rapid urbanization over the last 50 years, the majority of the population base (i.e. >90%) situated in all three countries live in urban centers. Table 1.1 lists the major cities and their populations within Brazil, Argentina and Paraguay that are located in the Paraná River Basin. Of the urban population of Paraguay, 40% live in or around the cities of Encarnación and

Table 1.1 Major urban centers located in the Paraná River Basin in the countries of Brazil, Argentina and Paraguay.

City	*Country*	*Tributary/Watershed*	*Population (millions)*
Asuncíon	Paraguay	Paraguay/Pilcomayo	0.53
Encarnacíon	Paraguay	Paraná	0.13
Ciudad del Este	Paraguay	Paraná	0.30
Formosa	Argentina	Paraguay	0.23
Resistencia	Argentina	Paraná	0.29
Posadas	Argentina	Paraná	0.35
Corrientes	Argentina	Paraná	0.4
Santa Fe	Argentina	Paraná/Salado	0.65
Rosario	Argentina	Paraná	1.16
São Paulo	Brazil	Tieté	10.66
Campinas	Brazil	Tieté watershed	1.10
Curitíba	Brazil	Iguazu	1.89
Foz do Iguaçu	Brazil	Iguazu/Paraná	0.25
Brasilia	Brazil	Paranaíba watershed	2.98
Goiâna	Brazil	Paranaíba watershed	1.30
Campo Grande	Brazil	Paraná watershed	0.77
Cuiabá	Brazil	Cuiabá	0.61

Ciudad del Este, and the other 60% live in or near the capital of Asunción. In Brazil, one third of the national population lives in urban centers located in the Paraná River Basin, including the cities and metropolitan areas of São Paulo, Brasilia, Goiâna and Curitíba (Table 1.1). Although the provinces in the northern part of Argentina are more sparsely populated than the more populous Buenos Aires province, there are still major urban centers located along the Paraná River, including Rosario, Corrientes and Santa Fe (Table 1.1). These cities contribute to a high demand for potable water, and water is also used for industrial processing and irrigation. At the same time, pollution from discharges of domestic and industrial wastewater and urban runoff contributed to the degradation of the water quality in the basin. Machado et al. (2014) tracked urban and industrial development in Curitíba along the Barigui River tributary of the Upper Paraná using the sedimentary record of contamination by polynuclear aromatic hydrocarbons (PAHs) and detected two depositional peaks during approximately 1910 and 1970; the latter attributed to the "Economic Miracle" period of development in Brazil during 1960 to 1980.

Within Brazil, Argentina and Paraguay, there is a strong economic and cultural connection with agriculture. However, since 1990, Argentina and Brazil have sharply increased agricultural output and have gained global market share for several major commodities, particularly soybeans (Schnepf et al., 2001). Brazil is today the world's largest producer of soybeans, coffee, cane sugar, orange juice, meat and tobacco, and agribusiness accounts for over 35% of its exports. Brazil is also the leading global producer of bioethanol. In Paraguay, approximately 55% of the area of the country is agricultural land, but a significant proportion of the land is used for subsistence farming. The arable land used for farming is concentrated in the eastern part of the country within the Paraná River Basin, and soybeans and cotton are the primary cash crops. In the northern provinces of Argentina, soybeans and corn are major cash crops and there are plans to expand the land area being farmed and increase rail service to carry agricultural products to market.

Deforestation and expansion in agricultural production has contributed to soil erosion, although more sustainable tillage practices have mitigated this somewhat (de Freitas and Landers, 2014; Poiretti and Domanski, 2014). Large amounts of fertilizers, pesticides and other agrochemicals are applied to these agricultural lands. Up until a few decades ago, wetlands in South America were relatively free of anthropogenic impacts (Junk et al., 2014). However, the expansion of the agricultural frontier due to the intensification of monoculture crops like soybeans, corn and sugar cane has displaced a high proportion of cattle production to the fluvial wetlands. For instance, expansion of pasture land for cattle grazing and the construction of water management infrastructure has resulted in the loss of wetlands in the Delta region of the Paraná River in Argentina (Magnano et al., 2018; Sica et al., 2016) and in the Pantanal wetland in Brazil (Junk and Nunes da Cunha, 2012).

The upper part of the Paraná River Basin that lies primarily within Brazilian territory has the largest hydroelectric potential in South America, but other parts of the basin have also been developed for hydroelectric power (Del Mar Rubio and Tafunell, 2014). Dam construction in this part of the basin began in 1895 with the relatively small Corumbatai Dam in the Tietê River. Beginning in the 1960s, the construction of hydroelectric power plants intensified, with the largest plants completed in 1974 (i.e. Jupiá), 1978 (i.e. São Simão), 1984 (i.e. Itaipu) and 1998 (i.e. Porto Primavera). There are several more dams in the planning stages (Stevaux et al., 2009). Hydroelectric dams are responsible for a range of environmental and sociological impacts. Dams modify and flood the landscape, impacting both terrestrial and aquatic ecosystems and changing sedimentation patterns (Shah and Kumar, 2008). They also disrupt or destroy local cultures and traditional ways of life, as well as inundate sites of historical and archeological importance (de Paiva Affonso et al., 2015).

In addition to the impacts of urbanization, agriculture and hydroelectric development on the Paraná River Basin, there are other potential problems from extraction of non-renewable resources that could affect the natural resources in the region. Brazil is among the top ten nations in the world for reserves of shale gas and much of these reserves lie within the Paraná River Basin (Lenhard et al., 2018). Access to these gas reserves will be mainly through hydraulic fracturing (i.e. "fracking") of the shale bedrock. This practice could have major environmental consequences, including the use of large volumes of water for the hydraulic fracturing and the potential for pollution of water resources from the chemicals in the fracturing fluids, as well as constituents from the gas and shale, such as methane, acetic acid, ammonia, chloride, metals and radioactive elements (Lenhard et al., 2018). These impacts could occur within surface waters, and also within the huge aquifer systems that lie beneath much of the Paraná River Basin (Sindico et al., 2018).

Ecosystem services

The subtitle of this book on the Paraná River Basin is 'Managing Water Resources to Sustain Ecosystem Services'. This title reflects the common theme running through the various chapters in this book related to how development within the basin has impacted ecosystem services, and the strategies for safeguarding these ecosystem services in order to sustain the growing human populations within the region. The Paraná River Basin is one of the greatest reserves of freshwater and one of the largest refuges for biodiversity on the planet. The riverine systems and wetlands within the basin provide water, sustenance and transportation corridors for Indigenous and traditional riverine communities. Urban centers of various sizes, including some of the largest cities in the Americas, depend on the natural resources and the ecological services of this region. In addition, there is an intimate link between fluvial discharge from the Paraná River into the Rio de la Plata estuary and the economies of the large population axes of Buenos Aires in Argentina and Montevideo in Uruguay.

The Paraná River Basin is of extraordinary ecological, social, cultural and economic value. The increasing pressure on the natural resources of the Paraná River Basin is of regional, national and international concern, given the unique characteristics of this watershed. Large development projects such as roads, dams and potentially hydraulic fracturing, as well as impacts on fisheries resources, have led to environmental degradation. Moreover, agriculture continues to expand, especially with the growth of soybean production and other monocultures. Vegetation cover is being lost due to urbanization and unsustainable agriculture practices, leading to erosion and sedimentation, runoff of agrochemicals and loss of habitat for native species. Industries and urban centers empty untreated or poorly treated wastewaters into this watershed.

There are many definitions of 'ecosystem services', but according to the Economics of Ecosystems and Biodiversity (TEEB) organization, ecosystem services are the direct and indirect contributions of ecosystems to human well-being, including to the quality of life and ultimately, human survival. The TEEB categorizes ecosystem services into four types:

- **Regulating services**: Benefits obtained from the regulation of ecosystem processes, such as climate regulation, natural hazard regulation, water purification and waste management and pest control.
- **Cultural services**: Non-material benefits that people obtain from ecosystems such as spiritual enrichment, recreation and aesthetic values.
- **Provision services**: Products obtained from ecosystems such as food, fresh water, wood, genetic resources and medicines.
- **Habitat services**: Providing habitat to migratory and non-migratory species and to maintain biodiversity and the viability of gene pools.

Regulating services

The development of hydroelectric dams and reservoirs has had both positive and negative impacts on the Paraná River Basin and the populations that live in the region, as discussed throughout this book. One of the positive benefits has been to regulate river flows within the Upper Reach of the basin; thereby ending the annual cycles of flooding that occurred prior to dam construction (Stevaux et al., 2009). In addition, the reservoirs provide a source of water for irrigation of agricultural lands, and an inland water transportation system (Tundisi, 2018). As discussed in this book, urbanization, industrial development, expansion of agriculture and aquaculture have put severe stresses on the capacity of the surface waters to assimilate discharges of waste and both urban and agricultural runoff, resulting in deteriorating water quality in many parts of the basin.

Cultural services

From a cultural perspective, the hydropower development, rapid urbanization and agricultural development in the basin have had both positive benefits and negative impacts. The growth in the economy has benefited large segments of society, but the unequal distribution of the wealth versus the social costs that come from this development is a source of conflict. For instance, there is a growing sense that the costs of the hydroelectric megaprojects in the Upper

Reach of the basin have been too great in terms of loss of cultural identity and damage to the natural environment (Aledo et al., 2015). On the other hand, tourism has developed as an important segment of the economy in several reservoirs within the basin. These cultural conflicts are illustrated by the current opposition to the development of the Paraguay-Paraná Waterway. This proposed water infrastructure project in the Paraguay River sub-basin would connect Brazil's interior with global shipping routes, but would have huge environmental impacts, including damage to the Pantanal region, as well as disruptions to traditional ways of life (Schulz et al., 2018).

Provision services

As discussed throughout the book, access to surface waters and groundwater in the region for irrigation and as a source of drinking water has had many positive benefits. However, unless regulations are put into place to protect both the quantity and quality of these aquatic resources, there will be a continued deterioration of the provisional services provided within the Paraná River Basin.

One of the provision services that is not discussed at length elsewhere in the book is the impacts of basin development on the traditional fishery, and the growth of aquaculture. The construction of hydroelectric dams in the Upper Reach of the Paraná has decimated the traditional fishery for native, migratory species that depend on lotic environments, such as "jaú" (*Zungaro zungaro*), "dourado" (*Salminus brasiliensis*) and "pintado" (*Pseudoplatystoma corruscans*), and now the dominant fish species, such "armado" (*Pterodorus granulosus*), "curimba" (*Prochilodus lineatuas*) and "curvina" (*Plagioscion squamosissimus*) thrive in the lentic environments provided by the reservoirs (Hoeinghaus et al., 2009). Cage aquaculture has been developed in the reservoirs with introduced species, such as Nile tilapia (*Oreochromis niloticus*) and sport fisheries have been created through the introduction of Cichlid species that are native to Amazonia (Britton and Orsi, 2012). While these non-native fish support provisional ecosystem services of considerable economic value, this must be balanced against the reduction in fish diversity and the loss of the cultural benefits of a traditional fishery. In parts of the basin that are undammed, such as the Paraguay River sub-basin, the traditional fishery is threatened by sedimentation and by runoff of contaminants from agriculture (Quirós et al., 2007). Finally, in order to support the traditional fishery within the Middle Reach of the Paraná River Basin, which generates total annual incomes estimated at USD $113 million, maintaining the annual flood cycles is key to providing the flood plain habitats that are necessary for reproduction and as larval nurseries (Rabuffetti et al., 2017).

Habitat loss and biodiversity

Habitat loss is one of the most pressing impacts on ecosystem services within the basin. This has resulted in loss of biodiversity and has caused several species to be placed on the IUCN "Red List" of endangered species. In the Espinal ecoregion in the northwest part of Argentina, loss of wetlands through expansion of agriculture has resulted in reduced biodiversity (Guida-Johnson and Zuleta, 2013). In the Delta region of the Lower Paraná, one third of the marshes have been lost due to forestry or by conversion to pasture lands (Sica et al., 2016). In the Upper Reach of the Paraná River Basin, the diversity of bird populations has been reduced as a result of the construction of dams that have reduced the extent of the flood plains (Almeida et al., 2017). In the Delta region, loss of wetlands as a result of expanded cattle farming has changed the bird assemblages (Sica et al., 2018). Agostinho et al. (2005) stated that pollution, eutrophication and hydroelectric development in southern Brazil were the reasons why 44 species of invertebrates and 134 fishes were on the list of threatened species in Brazil. Pelicice et al. (2016) provide a comprehensive review of the unsustainable activities in the Upper Paraná, such as hydropower development, water diversions, mining, aquaculture, agriculture and fishing, that have caused large scale losses of fish biodiversity. In the impounded regions of the Upper Paraná, small tributaries are the last preserves of early life stages of native fish, including several species at risk (da Silva et al., 2018).

Within the Paraná River Basin, a variety of species are threatened or endangered (Torrecilha et al., 2018; Agostinho et al., 2005; Garcia Fernandez et al., 1994). The current status of the *Piracanjuba* fish (*Brycon orbignyanus*), the Four-eyed frog (*Pleurodema kriegi*), the green kingfisher (*Chloroceryle americana*) and the Neotropical river otter (*Lontra longicaudis*) illustrate the ways in which multiple stressors, including habitat loss, climate change and the presence of invasive species can impact biodiversity within the basin. The Piracanjuba (Figure 1.2) is a neotropical migratory fish of great commercial interest in the Paraná River Basin. However, this fish species requires an area with preserved riparian forest for food and to complete its reproductive cycle. Because of the loss of riparian areas and overfishing, this species is currently classified as 'critically endangered' and is threatened with extinction in the Paraná basin (Ashikaga et al., 2015). The Four-eyed frog is listed as 'endangered' within the Paraná basin. Climate change is characterized

Figure 1.2 *Brycon orbignyanus*, known locally as 'Piracanjuba' is listed as a critically endangered fish species in the Paraná River Basin. The photograph was taken by David Morimoto and was provided by agreement with www.flickr.com.

by increasing mean temperatures, but it can also increase the frequency and intensity of extreme temperature events. There is evidence that extremes in seasonal temperatures caused by climate change are influencing the survival of the Four-eyed frog (Bacigalupe et al., 2018).

The green kingfisher or 'martin pescador verde' (Figure 1.3) is a piscivorous bird species threatened by loss of riparian habitat in the Paraná River Basin. Lozano and Malo (2013) described how kingfishers are particularly sensitive to the loss of riparian habitat in the Paraguay River sub-basin in Brazil. The range of the Neotropical river otter (Figure 1.4) extends through all parts of the Paraná basin, but its conservation status is listed by the IUCN as 'data deficient'. Gomez et al. (2014) studied the habitat of the river otter in the Delta region of the basin and they determined that the suitability of habitat was lower in areas with human settlements and high boat traffic. In a study of a population of the Neotropical river otter in Guatemala where their diet has changed as a result of the invasion of the armoured catfish (*Pterygoplichthys ambrosettii*), Juarez-Sanchez et al. (2019) reported that the breadth of the dietary niche and the trophic level of the otters was reduced because of their reliance on this fish species for food. The armoured catfish is also an invasive species within the Paraná River Basin (Noble et al., 2018). It is possible that an over-reliance on this invasive fish in the diet of the river otters is an added challenge to the survival of the population in the basin. This points up the importance of invasive species as an ecological stressor that may contribute to the loss of biodiversity.

Figure 1.3 The green kingfisher or 'martin pescador verde' (*Chloroceryle americana*) is a bird species threatened by loss of riparian habitat in the Paraná River Basin. The photograph was provided by agreement with www.flickr.com.

Figure 1.4 This Neotropical river otter (*Lontra longicaudis*) is distributed throughout the Parana River Basin and is threatened by habitat loss and other stressors. The photograph was provided by agreement with www.shutterstock.com.

In addition to the armoured catfish, there are several other invasive species that have spread throughout the Paraná River Basin. The usual ways in which these species are introduced is by release in the ballast water from vessels or by intentional or accidental release by aquarium owners. Like many watersheds in Europe and the Americas, parts of the Paraná basin have been invaded by the Asian clam, *Corbicula fluminea* (Figure 1.5). Studies conducted in the Paraná and other river basins around the world where this exotic species has invaded have shown that this benthic clam can change the ecological characteristics of lotic and lentic environments. Boltovskoy et al. (1995) showed that high densities of the clams in the Paraná River were associated with reductions in phytoplankton in the water column. In a study conducted in Portugal, Sousa et al. (2008) showed that Asian clam populations sequestered a high proportion of the carbon available for benthic production and thus altered the ecosystem services of a river basin.

Summary of book chapters

In this book, experts on the various physical, chemical, biological and ecological factors that influence the ecological functioning of the Paraná River Basin will provide in-depth descriptions of the past, current and future challenges for managing this important hydrological system. Authors contributing to other chapters in the book will describe the water governance framework within the basin and both the challenges and the potential solutions for managing water resources.

Figure 1.5 The Asian clam (*Corbicula fluminea*) which is an invasive species inhabiting the Paraná River Basin and other river basins around the world. The source of this photograph was the Smithsonian Environmental Research Center and it was provided through an agreement with creativecommons.org.

Chapter 2 provides an overview of the physiography, climatology and hydrology of the Paraná River Basin. As described in this chapter, climatological, geological and hydrological parameters, as well as human influences are affecting both the quality and quantity of surface waters in the basin. Additionally, these surface waters recharge the aquifers of the basin, including the Guarani Aquifer, one of the largest continental groundwater reservoirs in the world. The authors of this chapter of the book provide a critical analysis of the threats to water resources within the Paraná River Basin caused by population growth and urbanization, water use by industry and agriculture and hydroelectric development. This chapter includes a historical perspective on the experiences within the basin related to the management of shared waters. The geomorphology and sedimentation characteristics within the basin are described in Chapter 3. The Paraguay River and its tributaries contribute large amounts of sediment to the system as a result of the transport of silt and sand from the Andean region in the eastern part of Bolivia. The Upper Reach of the Paraná River has a relatively low sediment load as a result of construction of a series of reservoirs behind hydroelectric dams. The history of dam development in the basin is described in Chapter 4, as well as the impacts of this development on human populations, aquatic ecosystems, water flows and water quality. The authors of the chapter argue that integrated management of watersheds and reservoirs is a fundamental requirement to ensure sustainability within the basin.

The authors of Chapter 5 in the book describe the great diversity of aquatic ecosystems in the basin, including ecologically important wetlands, and they also describe the impacts of development on these ecosystems. Also included in this chapter is a description of the impacts of La Niña-El Niño climate oscillations on natural ecosystems. Of course, long-term changes in climate may have huge consequences for the basin. The authors of Chapter 6 of the book used Global Climate Models to make projections about the changes in temperature, rainfall and discharge that will occur in different parts of the basin as a result of climate change.

Temperatures are predicted to rise in the whole basin, while changes in precipitation will vary across the different regions. The models predict that rivers that flow from north to south within the Paraná headwaters, which are mostly within Brazil, will face reductions in discharge, and the same trend is predicted for rivers that flow from the western part of the basin. The authors of Chapter 7 review the available literature describing water quality within the basin, including the loads of suspended sediments and the concentrations in water and sediments of metals from both anthropogenic and geogenic sources. The authors of this chapter also review the literature on the concentrations in water and sediments of aromatic hydrocarbons and pesticides that are currently used or were used in the recent past.

Chapter 8 describes the current regime for transboundary governance of water resources and hydroelectric power within the Paraná River Basin, which is characterized by multiple bilateral and multilateral legal instruments, and by Commissions that provide the institutional backbone for the governance regime. While there is a highly fragmented governance system within the basin, the authors of this chapter argue that no major conflicts have occurred throughout the course of the management of the Paraná River and its tributaries. Chapter 9 describes the current state of progress in meeting Sustainable Development Goal 6 for providing access to clean water and sanitation for all. The national data generated for Argentina, Brazil and Paraguay indicate that there will be significant challenges in addressing the target to provide universal access to safe drinking water and to provide facilities for sanitation and hygiene, particularly among the urban and rural poor.

Acknowledgments

We appreciate the assistance of Hamid Mehmood at UNU-INWEH for providing geospatial data for the Paraná River Basin in ArcMap format. We thank Erin Hayward at Trent University for preparing the basin maps from the geospatial data for this chapter and for other chapters in the book. We also thank Rob Loney at Trent University for his assistance in formatting figures and photographs.

References

Agostinho, A.A., Thomaz, S.M., Gomes, L.C. 2005. Conservation of the biodiversity of Brazil's inland waters. *Conservation Biology* 19:646–652.

Aledo, A., Garcia-Andreu, H., Pinese, J. 2015. Using causal maps to support ex-post assessment of social impacts of dams. *Environmental Impact Assessment Review* 55:84–97.

Almeida, B.A., Gimenez, M.R., Dos Anjos, L. 2017. Wading bird functional diversity in a floodplain: Influence of habitat types and hydrological cycle. *Austral Ecology* 42:84–93.

Ashikaga, F.Y., Orsi, M.L., Oliveira, C., Senhorini, J.A., Foresti, F. 2015. The endangered species *Brycon orbignyanus*: Genetic analysis and definition of priority areas for conservation. *Environmental Biology of Fishes* 98:1845–1855.

Ashworth, P.J., Lewin, J. 2012. How do big rivers come to be different? *Earth Science Review* 114:84–107.

Bacigalupe, L.D., Gaitán-Espitia, J.D., Barria, A.M., Gonzalez Mendez, A., Ruiz-Aravena, M., Trinder, M., Sinervo, B. 2018. Natural selection on plasticity of thermal traits in a highly seasonal environment. *Evolutionary Applications* 11:2004–2013.

Barletta, M., Jaureguizar, A.J., Baigun, C., Fontoura, N.F., Agostinho, A.A., Almeida-Val, V.M.F., Val, A.L., Torres, R.A., Jimenes-Segura, L.F., Giarrizzo, T. 2010. Fish and aquatic habitat conservation in South America: A continental overview with emphasis on neotropical systems. *Journal of Fish Biology* 76:2118–2176.

Blanc, J. 2018. Itaipu's forgotten history: The 1965 Brazil-Paraguay border crisis and the new geopolitics of the Southern Cone. *Journal of Latin American Studies* 50:383–409.

Boltovskoy, D., Izaguirre, I., Correa, N. 1995. Feeding selectivity of *Corbicula fluminea* (Bivalvia) on natural phytoplankton. *Hydrobiologia* 312:171–182.

Britton, J.R., Orsi, M.L. 2012. Non-native fish in aquaculture and sportfishing in Brazil: Economic benefits versus risks to fish diversity in the upper River Paraná Basin. *Reviews in Fish Biology and Fisheries* 22:555–565.

Carol, E., Mas-Pla, J., Kruse, E. 2013. Interaction between continental and estuarine waters in the wetlands of the northern coastal plain of Samborombón Bay, Argentina. *Applied Geochemistry* 34:152–163.

da Silva, P.S., Miranda, L.E., Makrakis, S., de Assumpção, L., Días, J.H.P., Makrakis, M.C. 2018. Tributaries as biodiversity preserves: An ichthyoplankton perspective from the severely impounded Upper Paraná River. *Aquatic Conservation: Marine and Freshwater Ecosystems* 29:258–269.

de Freitas, P.L., Landers, J.N. 2014. The transformation of agriculture in Brazil through development and adoption of zero-tillage conservation agriculture. *International Soil and Water Conservation Research* 2:35–46.

de Paiva Affonso, I., Azevedo, R.F., Carneiro Dos Santos, N.L., Dias, R.M., Agostinho, A.A., Gomes, L.C. 2015. Pulling the plug: Strategies to preclude expansion of dams in Brazilian rivers with high priority for conservation. *Nature Conservation* 13:199–203.

Del Mar Rubio, M., Tafunell, X. 2014. Latin American hydropower: A century of uneven evolution. *Renewable and Sustainable Energy Review* 38:323–334.

Drago, E.C., Amsler, M.L. 1998. Bed sediment characteristics in the Paraná and Paraguay rivers. *Water International* 23:174–183.

Garcia Fernandez, J.J., Ojeda, R.A., Fraga, R.M., Diaz, G.B., Baigun, R.J. 1994. *Mamiferos y Aves Amenazados de la Argentina*. Fundación para Conservación de las Especies y Medio Ambiente, Argentina, p. 221.

Gomez, J.J., Tunez, J.I., Fracassi, N., Cassin, M.H. 2014. Habitat suitability and anthropogenic correlates of Neotropical river otter (*Lontra longicaudis*) distribution. *Journal of Mammalogy* 95:824–833.

Guida-Johnson, B., Zuleta, G.A. 2013. Land-use land-cover change and ecosystem loss in the Espinal ecoregion, Argentina. *Agriculture, Ecosystems and Environment* 181:31–40.

Hoeinghaus, D.J., Agostinho, A.A., Gomes, L.C., Pelicice, F., Okada, E.K., Latini, J.D., Kashiaqui, E.A.L., Winemiller, K.O. 2009. Effects of river impoundments on ecosystem services of large tropical rivers: Embodied energy and market value of artisanal fisheries. *Conservation Biology* 23:122–1231.

Iriondo, M.H. 2007. Geomorphology. In: Iriondo, M.H., Paggi, J.C., Parma, M.J. (eds.), *The Middle Paraná River*. Springer-Verlag, Berlin, pp. 33–52.

Isupova, M.V., Mikhailov, V.N. 2018. Long-term variations of water runoff and suspended sediment yield in the Paraná and Uruguay rivers. *Water Resources* 45:846–860.

Juarez-Sanchez, D., Blake, J.G., Hellgren, E.C. 2019. Variation in Neotropical river otter (*Lontra longicaudis*) diet: Effects of an invasive prey species. *PLoS ONE* 14(10):e0217727. DOI:10.1371/journal.pone.0217727

Junk, W.J., Nunes da Cunha, C. 2012. Pasture clearing from invasive woody plants in the Pantanal: A tool for sustainable management or environmental destruction? *Wetland Ecology and Management* 20:111–122.

Junk, W.J., Nunes da Cunha, C., Wantzen, K.M., Petermann, P., Strüsmann, C., Marques, M.I., Adis, J. 2006. Biodiversity and its conservation in the Pantanal of Mato Grosso, Brazil. *Aquatic Science* 68:278–309.

Junk, W.J., Peidade, M.T.F., Lourival, R., Whittmann, F., Kandus, P., Lacwerda, L.D., Bozelli, R.L., Esteves, F.A., Nunes da Cunha, C., Maltchik, L., Schönggart, J., Scaeffer-Novelli, Y., Agostinho, A.A. 2014. Brazilian wetlands: Their definition, delineation, and classification for research, sustainable management, and protection. *Aquatic Conservation of Marine and Freshwater Ecosystems* 24:5–22.

Latrubesse, E.M., Abad, J.D. 2018. Preface to the volume, Large Rivers. *Geomorphology* 302:1–2.

Lenhard, L.G., Andersen, S.M., Coimbra-Araújo, C.H. 2018. Energy-environmental implications of shale gas exploration in Paraná hydrological basin, Brazil. *Renewable and Sustainable Energy Review* 90:56–69.

Lozano, J., Malo, A.F. 2013. Relationships between human activity and richness and abundance of some bird species in the Paraguay River (Pantanal, Brazil). *Andeola* 60:99–112.

Machado, K.S., Figueira, R.C.L., Côcco, L.C., Froehner, S., Fernandes, C.V.S., Ferreira, P.V.L. 2014. Sedimentary record of PAHs in the Bariguí River and its relation to the socioeconomic development of Curitíba, Brazil. *Science of the Total Environment* 482–483:42–52.

Magnano, A.L., Nanni, A.S., Krug, P., Astrada, E., Vicari, R., Quintana, R.D. 2018. Effects of livestock exclusion on density, survival and biomass of the perennial sagebrush grass *Hymenachne pernambucense* (Poaceae) from a temperate fluvial wetland. *Acta Oecologica* 86:72–78.

Nakabashi, L., Goncalves Pereira, A.E., Sachsida, A. 2013. Institutions and growth: A developing country case study. *Journal of Economic Studies* 40:614–634.

Noble, A.B., Lima, F.P., Freitas-Souza, D., Queiroz, J., Garcia, D.A., Orsi, M.L., Vidotto-Magnoni, A.P. 2018. Distribution of non-native suckermouth armoured catfish *Pterygoplichthys ambrosettii* in the upper Paraná River basin, Brazil. *Journal of Fish Biology* 92:1198–1206.

Orfeo, O., Stevaux, J. 2002. Hydraulic and morphological characteristics of middle and upper reaches of the Paraná River (Argentina and Brazil). *Geomorphology* 44:309–322.

Pelicice, F.M., Azevedo-Santos, V.M., Vitule, J.R., Orsi, M.L., Lima, D.P., Jr., Magalhães, A.L.B., Pompeu, P.S., Petrere, M., Jr., Agostinho, A.A. 2016. Neotropical freshwater fishes imperilled by unsustainable policies. *Fish and Fisheries* 18:1119–1133.

Poiretti, R., Domanski, J. 2014. The transformation of agriculture in Argentina through soil conservation. *International Soil and Water Conservation Research* 2:11–20.

Quirós, R., Bechara, J.A., de Resende, E.K. 2007. Fish diversity and ecology, habitats and fisheries for the un-dammed riverine axis Paraguay-Parana-Rio de la Plata (Southern South America). *Aquatic Ecosystem Health Management* 10:187–200.

Rabuffetti, A.P., Gorski, K., Espinola, A., Abrial, A., Amsler, M.L., Paira, A.R. 2017. Long-term hydrologic variability in a large subtropical floodplain river; Effects on commercial fisheries. *River Research Applications* 33:353–363.

Schnepf, R.D., Dohlman, E., Bolling, C. 2001. *Agriculture in Brazil and Argentina: Developments and Prospects for Major Field Crops*. Economic Research Service, U.S. Department of Agriculture, Agriculture and Trade Report, WRS-01-3, Washington, D.C.

Schulz, C., Martin-Ortega, J., Glenk, K. 2018. Value landscapes and their impact on public water policy preferences. *Global Environmental Change* 53:209–225.

Shah, Z., Kumar, M.D. 2008. In the midst of the large dam controversy: Objectives, criteria for assessing large water storages in the developing world. *Water Resource Management* 22:1799–1824.

Sica, Y.V., Gregorio, I., Gavier-Pizzaro, G., Pidgeon, A.M., Travaini, A., Bustamente, J., Radeloff, V.C., Quintana, R.D. 2018. Changes in bird assemblages in a wetland ecosystem after 14 years of intensified cattle farming. *Austral Ecology* 43:786–797.

Sica, Y.V., Quintana, R.D., Radeloff, V.C., Gavier-Pizarro, G.I. 2016. Wetland loss due to land use change in the Lower Paraná River Delta, Argentina. *Science of the Total Environment* 568:967–978.

Sindico, F., Hirata, R., Manganellie, A. 2018. The Guarani aquifer system: From a beacon of hope to a question mark in the governance of transboundary aquifers. *Journal of Hydrology: Regional Studies* 20:49–59.

Sousa, R., Nogueira, A.J.A., Gaspar, M.B., Antunes, C., Guilhermino, L. 2008. Growth and extremely high production of the non-indigenous invasive species *Corbicula fluminea* (Müller 1774): Possible implications for ecosystem functioning. *Estuarine, Coastal and Shelf Science* 80:289–295.

Stevaux, J.C., Martins, D.P., Meurer, M. 2009. Changes in a large regulated tropical river: The Paraná River downstream from the Porto Primavera Dam, Brazil. *Geomorphology* 113:230–238.

Torrecilha, S., Ochoa-Quintero, J.M., Laps, R.R., Ribeiro, D.B. 2018. Records of threatened bird and mammal species in Matto Grosso do Sul State, Brazil. *Data in Brief* 17:1326–1330.

Tundisi, J.G. 2018. Reservoirs: New challenges for ecosystem studies and environmental management. *Water Security* 4–5:1–7.

Vega, L.F., Nunes da Cunha, C., Rothaupt, K.-O., Moreira, M.Z., Wantzen, K.M. 2014. Does flood pulsing act as a switch to store or release sediment-bound carbon in seasonal floodplain lakes? Case study from the Colombian Orinoco-Llanos and the Brazilian Pantanal. *Wetlands* 34:177–187.

2

Water resources

Status and management challenges

Paulo Cesar Rocha, Antonio Cezar Leal, Renata Ribeiro de Araújo, Cláudio Antonio Di Mauro, Wagner Costa Ribeiro, and Hamid Mehmood

Key messages

- The basin of the Paraná River, with a total area of 2,605,000 km^2 is shared between Brazil, Argentina, Paraguay and Bolivia, with close to half of the basin area located in Brazil. The mean annual river flow into the Atlantic Ocean is around 560 km^3, which places the river within the five largest in the world by flow volume.
- Rainfall is abundant in the eastern part of the basin, particularly in the Upper Paraná, but more limited in the western parts of the basin. Similarly, variability of river flow between calendar months is relatively low in the eastern and mainstream regions of the Paraná, but there is greater variability in the western part of the basin.

- Water challenges in the Paraná River Basin are associated with population growth and urbanization, increasing water use for industry and agriculture, deterioration of water quality caused by the discharges of urban sewage and industrial effluents, intensive livestock farming, and the impacts of runoff of agrochemicals in regions where there is extensive cultivation of soybeans, sugar cane and other monoculture crops.
- The Intergovernmental Coordinating Committee of the Countries of the La Plata Basin (CIC) has identified 11 Cross-Border Critical Issues, which include: (1) Extreme hydrological events; (2) Water quality degradation; (3) Sedimentation of waterways and bodies of water; (4) Alteration and loss of biodiversity; (5) Unsustainable use of fishery resources; (6) Unsustainable use of aquifers in critical areas; (7) Water use conflicts and the environmental impact of crop irrigation; (8) Lack of disaster contingency plans; (9) Poor state of waterborne health parameters and deterioration of sanitation; (10) Navigational limitations, (11) Development of hydropower potential.
- To address these critical issues and other water management problems in the basin, several agreements have been established between the four countries that share the basin (i.e. Brazil, Argentina, Paraguay, Bolivia) and management bodies have been formed to execute the articles of the agreements.
- There are current examples of water resource management in the Paraná River Basin, including governmental (i.e. top down) and local (i.e. bottom up) approaches. The four countries that share the Paraná River Basin currently have the institutional and legal frameworks, principles, systems and instruments for water management that are based on a participatory model. Strengthening of River Basin Committees is crucial for decision-making and addressing major basin challenges.

Basin physiography and hydrology

The Paraná Basin covers an area of 2,605,000 km^2, of which 1,260,000 km^2 is within the territory of Brazil, 730,000 km^2 is in Argentina, 410,000 km^2 is in Paraguay and only 205,000 km^2 is located in the eastern margin of Bolivia. The river originates at the confluence of the Paranaíba and Grande rivers at an elevation of around 1,100 meters above sea level (masl) and eventually drains into the Rio de la Plata estuary. The total length of the river from the Paranaiba source to the La Plata mouth is 3,740 km (www.fao.org/nr/water/aquastat/basins/la-plata/print1.stm)

The topography of the basin downstream of the confluence of the Paraguay and Paraná rivers is mostly flat (Figure 2.1). However, the Upper stretch of the Paraná River is characterized by waterfalls and rapids as a result of the geological structure of tiered highlands (Figure 2.1). Therefore, this part of the basin has been extensively developed for the generation of hydroelectric power. The Paraguay River, which is the main tributary of Paraná, with an area of almost 1,100,000 km^2 and a length of over 2,500 km, is shared by Brazil, Argentina and Bolivia (Figure 2.1). Located in the northern part of the Paraguay sub-basin is one of the most well-known and largest wetlands in the world, the Pantanal, which extends over an area of approximately 60,000 km^2 (Figure 2.2). As illustrated in Figure 2.2, land use in the eastern and southern parts of the basin is dominated by agriculture (i.e. croplands), while forested areas cover much of the western part of the basin.

The western margin of the Paraguay River sub-basin originates in the Andes Mountains in the eastern part of Bolivia (Figure 2.1). Because of the alluvial nature of this western watershed, an enormous volume of suspended solids material is carried by the Verde, Pilcomayo and Bermejo rivers into the Paraguay River. Although the area of the Paraguay River sub-basin is

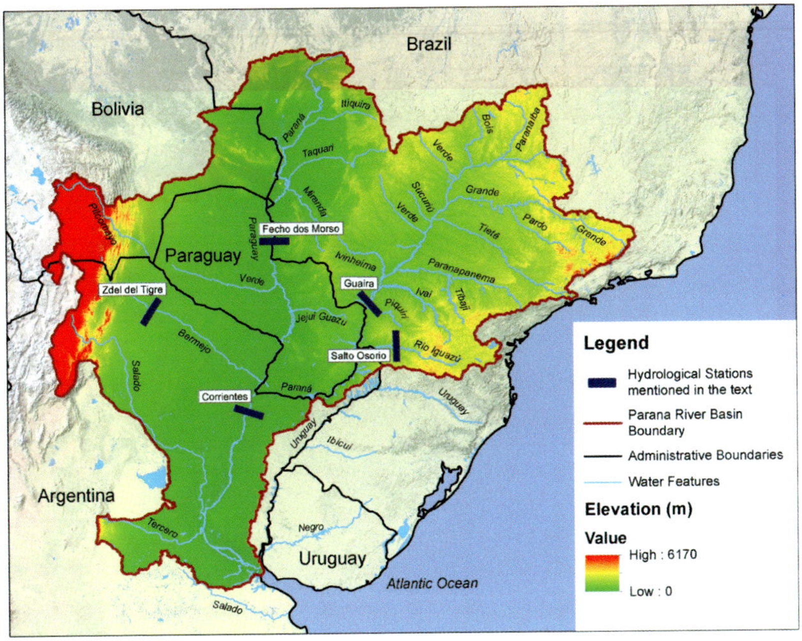

Figure 2.1 A topographic map of the Paraná River Basin. The five hydrological stations for which annual flow data are presented in Figure 2.8 are labelled.

Source: Based on GTOPO 30 data of US Geological Survey's EROS Data Center.

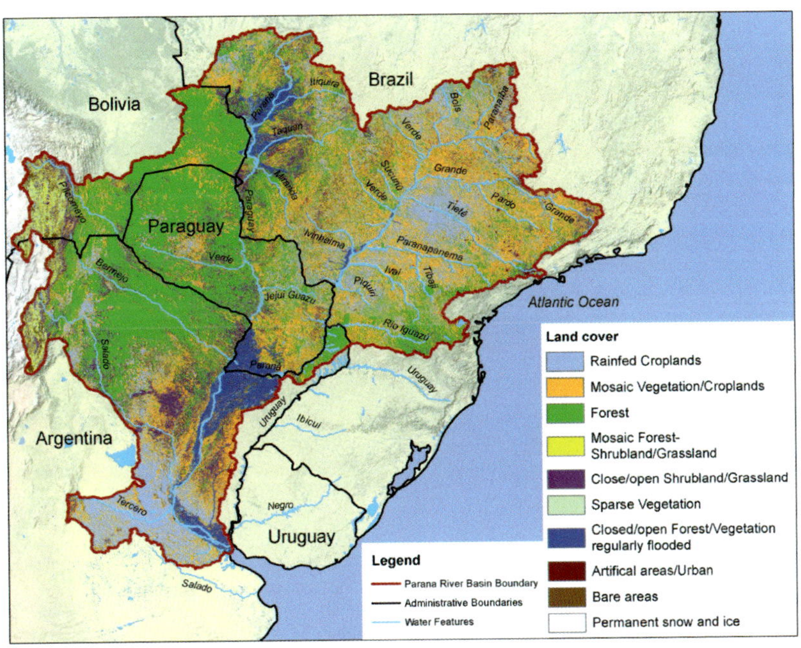

Figure 2.2 Land cover of the Paraná River Basin.

Source: Based on GlobCover 2009 data of European Space Agency.

large, the river flow is relatively low because of the dry climate in the western part of the basin (Collischonn et al., 2016). The backflow produced at the confluence with the waters of the Paraná River causes regular flooding and contributes to the meandering course of this part of the Paraguay River (Figure 2.3). Downstream of the confluence, in the Middle Reach of the basin, there are extensive flood plains (Figure 2.2). Figure 2.4 illustrates the complexity of the flood plain in this part of the basin.

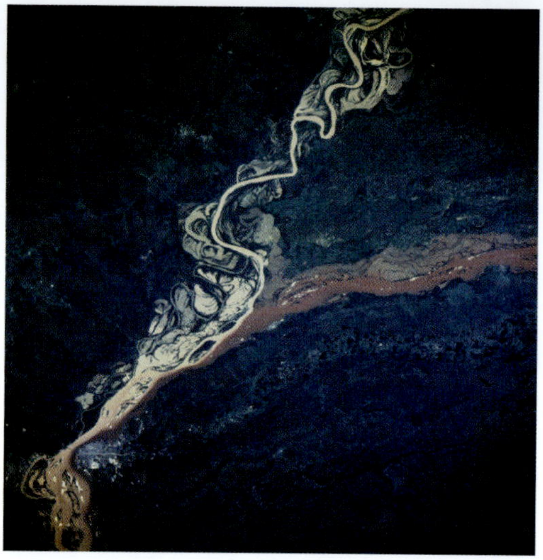

Figure 2.3 Satellite image of the confluence of the Paraguay River (top) and the Paraná River (right), upstream of the city of Corrientes.

Source: Photograph from the NASA Goddard Space Flight Center obtained through an agreement with Flickr.com.

Figure 2.4 A satellite image of the alluvial flood plain in the Middle Reach of the Paraná River Basin.

Source: Photograph from the NASA Goddard Space Flight Center obtained through an agreement with Flickr.com.

The Lower reach of the basin begins downstream of the city of Rosario in Argentina, at the most upstream influence from the tides that originate in the Rio de la Plata estuary. The delta of the Paraná River begins around 320 km from the river mouth and covers an area of over 14,000 km^2 (Meis and Llano, 2019). As illustrated in Figure 2.5, the discharge from Paraná River, as well as the adjacent Uruguay River basin is a major contributor to the sediment flux into the Atlantic Ocean. Although the complexity of the Paraná delta makes it difficult to accurately estimate the river flow, some recent estimates indicate that it is around 17,700 m^3 s^{-1}, as measured over the 1991 to 2013 period (Meis and Llano, 2019), which converts to almost 560 km^3 a year in volumetric units, and places the Paraná River among the top five largest rivers in the world.

The climate within the Paraná River Basin varies from tropical in the north to subtropical in the Middle Reach and temperate in the south. The mean annual temperature ranges from around 15°C in the south to more than 25°C in the northwest (Figure 2.6). The eastern part of the basin, which is mostly located in Brazil, receives more rainfall than the western part (Figure 2.7). The annual rainfall in the upper reaches of the basin can be as high as 2,000 mm, while the annual rainfall in the dry Chaco region west of the Paraguay River can be as low as 300 mm (Barros et al., 2015). This disparity in rainfall explains why the Paraná River and the Paraguay River are very dissimilar in terms of discharge. The contribution of the Paraguay River sub-basin is only around 20% of the average flow of Paraná River at the city of Corrientes (Meis and Llano, 2019).

Figure 2.5 A satellite image of the Paraná River delta and the Rio de la Plata estuary showing the extent of sediment transport into the Atlantic Ocean.

Source: Photo credit to Jacques Descloitres, MODIS Rapid Response Team, NASA (Public domain). commons.wikimedia.org.

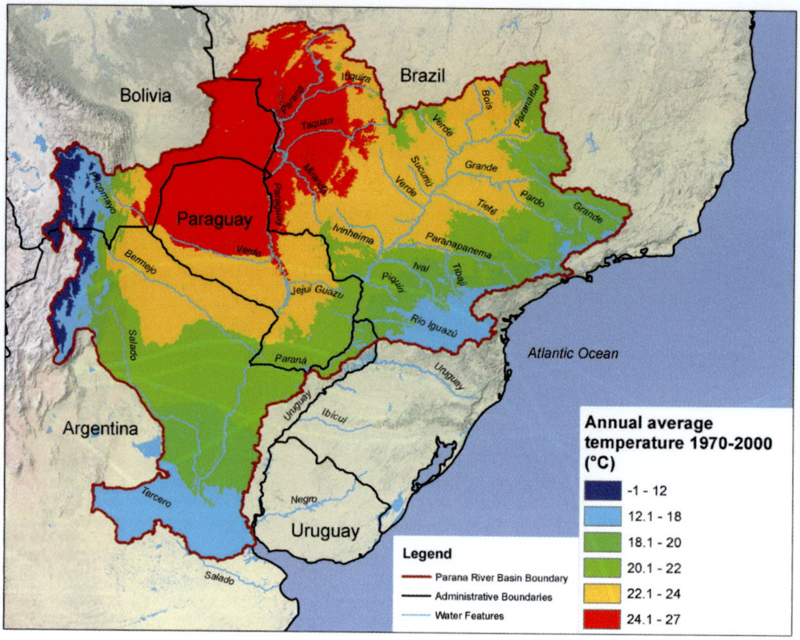

Figure 2.6 Mean annual air temperatures in the Paraná River Basin.

Source: Based on the WorldClim Data (http://worldclim.org/version2), with 2.5' spatial resolution.

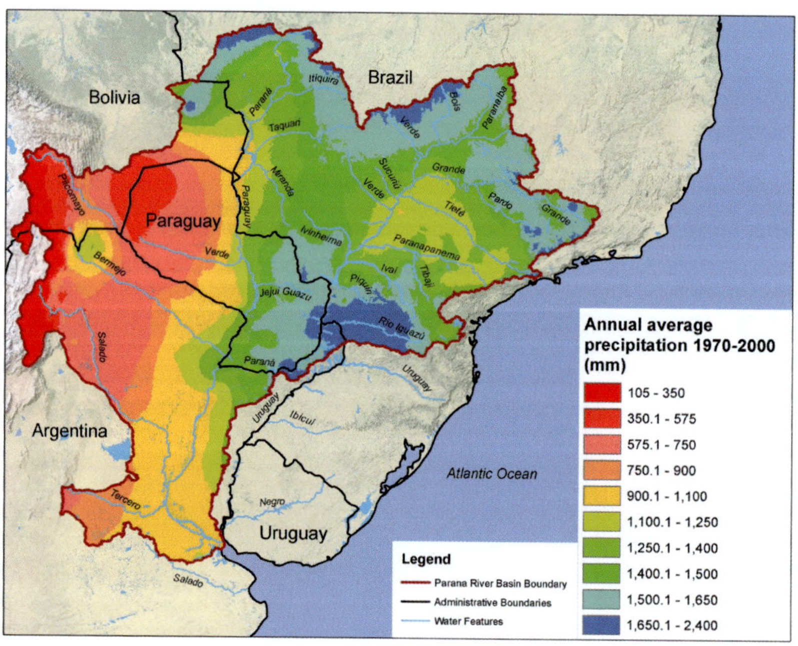

Figure 2.7 Mean annual precipitation (mm) in the Paraná River Basin.

Source: Based on the WorldClim Data (http://worldclim.org/version2), with 2.5' spatial resolution

Figure 2.7 also shows that there is a north-south gradient in precipitation, with the Middle and Lower Reaches of the basin being drier than the upstream parts. In addition, El Niño-La Niña events have a strong influence on the extent and severity of floods and droughts (Krepper and Venturini, 2009), as well as the flows throughout the basin (Puig et al., 2016).

Based on records over the period 1909 to 2013 at downstream stations below the confluence with Paraguay River (i.e. Corrientes, Timbues), Meis and Llano (2019) concluded that flows in the Paraná River have not decreased but rather have increased slightly over the last 20–30 years. Normalized seasonal flow distributions at different parts of the basin illustrated in Figure 2.8 indicate that flow regimes are rather stable throughout a year, with exception of flows in some western tributaries (i.e. Bermejo River), due to a more variable rainfall regime in those areas.

There are several aquifer systems in the Paraná River Basin, the largest and the most important being the Guarani Aquifer System (GAS), shared by Argentina, Brazil, Paraguay and Uruguay. The GAS is estimated to cover an area of approximately 1.2 million km^2 (Amore, 2011). Water is contained in porous sandstone and is confined is some areas by basalt rock. The extraction of water from the aquifer typically requires wells with depths ranging from 200 to 1,000 m. In general, the water is of good quality in shallower regions of the GAS. However, in some deep wells, the water is brackish and unfit for human consumption (Brião et al., 2014). In some regions of Argentina, groundwater from the GAS is only used for thermal baths because

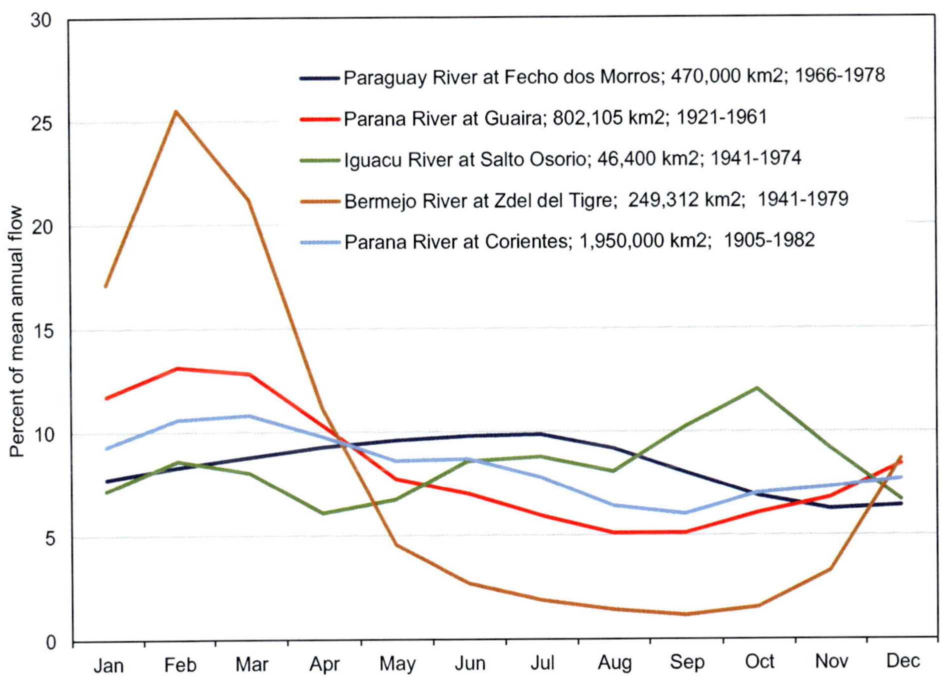

Figure 2.8 Seasonal flows expressed as percent of mean annual flow at selected locations within the Paraná River Basin. The data are based on observed flow data recorded at the five hydrological gauging stations shown in Figure 2.1.

Source: Based on UNESCO data: https://wayback.archive-it.org/10611/20160803175443/http://webworld.unesco.org/water/ihp/db/shiklomanov/index.shtml

of a high content of salt, sulfates, total dissolved solids and fluoride (Brião et al., 2014). Urban pollution and intense exploitation for irrigation are currently the most critical issues for groundwater management (Amore, 2011). There is growing concern that future exploitation of shale gas resources in the region may contaminate groundwater (Lenhard et al., 2018).

Water resource challenges

Most of the rivers and tributaries in the Paraná River Basin (PRB) are influenced by stressors that have modified and continue to modify the quantity and quality of water. These stressors and their consequences are of a transboundary, basin-wide nature. Water management in the PRB should be understood in the broader context of the La Plata Basin, with its history of agreements and the formation of governing organizations, as well as the national and governmental priorities of the countries of Argentina, Bolivia, Brazil and Paraguay. The Intergovernmental Coordinating Committee of the Countries of the La Plata Basin (CIC) identified 11 Cross-Border Critical Issues (CCIs), noting their main causes and presenting recommendations for resolution and management, which are within the scope of this chapter of the book. To address these CCIs and other specific problems in the sub-basin, several agreements have been established between the countries and management bodies have been formed to execute the articles of the agreements (Table 2.1). The main causes and the approaches used to manage these critical problems are the focus of this chapter. These drivers of change will

Table 2.1 Current and emerging cross-border critical issues (CCIs) in the La Plata Basin

CCI	*Causes*	*Location*
Extreme hydrological events (floods and droughts)	Lack of urban and territorial planning; poor coordination of information on extreme events, and lack of regional disaster prevention policies; lack of education and awareness.	Floods in the Apa, Iguazu, Paraguay, Pilcomayo, Bermejo and Paraná rivers
Water quality degradation	Inadequate treatment of wastewater; lack of training for environmental managers; lack of policies that encourage the use of clean technology and waste minimization; the lack of compliance with existing regulations.	The entire PRB
Sedimentation of waterways and bodies of water	Inadequate use and management of land (expanding agricultural activity, use of marginal soils, elimination of natural pastures, overgrazing); lack of incentives, policies and training to apply sustainable agricultural techniques by state agencies.	The entire PRB
Alteration and loss of biodiversity	Replacement of natural ecosystems with productive activities; lack of incentives for the care and conservation of natural systems; lack of protocols for the control of invasive species; lack of social awareness about the value of water resources and biodiversity	The entire PRB

Unsustainable use of fisheries resources	Overexploitation of species of commercial interest; the lack of technical and political coherence in the design and implementation of fishery policies; the lack of harmonious and integrated policies for the protection of aquatic life throughout the basin; The use of unsustainable techniques and difficulty in accepting new technologies.	The entire PRB
Unsustainable use of aquifers in critical areas	Contamination of aquifers from agriculture and domestic and industrial discharges; the lack of groundwater use management; the lack of transboundary institutional coordination for shared control and management, and low social participation.	Guaraní; Yrenda (PY), Toba (AR), Tarijeño (BO), Río Apa aquifer (BR and PY) and Pantanal, Furnas, Caiuá, Parecis and Serra Geral (BR)
Water use conflicts and environmental impact of crop irrigation	Scarce or poor information available on shared water resources (inventory of uses and availability); the lack of joint management bodies for shared water resources; asymmetries in legal-institutional structures for the integrated management of shared resources; a lack of social awareness about the value of resources and their limited availability.	All of the PRB, and in particular the rivers Cuareim, Pilcomayo, Tietê and Paraná. The rice basins and the lower basin of the Paraná-Paraguay
Lack of disaster contingency plans	Risk of dam breakages due to errors in dam operations; lack of review of the safety criteria for dams, taking into account the impacts of climate change; the absence of national and transnational regulations governing dam safety; the lack of awareness surrounding the risks to the populations located downstream of these types of installations and to the operating companies themselves.	Rivers: Paraná, Iguazú, Negro and Pilcomayo
Poor waterborne health parameters and deterioration of environmental sanitation	The lack of information on waterborne diseases; inefficiency of control over industrial discharges and agrochemicals; the asymmetry of legal and technical criteria for the management of water resources and public health, and resistance to change in habits.	The entire PRB
Navigational limitations	The lack of infrastructure to overcome natural critical points; inadequate joint institutional management; asymmetries and weaknesses in country regulations; and the preference for transport by land.	Paraná-Paraguay Waterway (AR, BO, BR, PY and Paraná and Tietê (BR, PY)
Development of hydropower potential	An analysis of the threats has not been carried out	

Note: R = Argentina; BO = Bolivia; BR = Brazil; PY = Paraguay.

Source: CIC (2017a, 2017e) and CIC (2017b).

surely intensify in the future as the basin countries continue to expand their agricultural and industrial development and provide services that improve the living standards of their growing populations (OAS, 2005).

Impacts of development on water resources

The primary developments that have had impacts on the environment within the PRB include: (i) development of the many hydropower dams and reservoirs in the Upper Paraná River in Brazil over the period from 1960 to 1990; (ii) deforestation in the Paraná and Paraguay basins from 1950 to 1990; (iii) introduction of intensive agricultural practices after 1970; (iv) urban development, with changes to flood regimes, and (v) navigation and conservation efforts in the Upper Paraguay River (Tucci, 2004). The presence of abundant mineral resources, forests and fertile soils for agriculture favors economic development and makes the PRB an attractive area for population settlement within the territories of Brazil and Argentina, with a strong agricultural and livestock component and a significant level of industrial production and urban services. In Paraguay, the economy is primarily based on agriculture (OAS/DSD, 2019).

In this context, this chapter covers the challenges for the integrated and participatory management of the PRB among the countries that share the waters within the basin; that is, Argentina, Brazil, Paraguay and Bolivia. Each of these countries has water resource policies and water resource management systems based on a systemic-participatory model, with decentralization, participation and integration, and with different levels of implementation. Although there are many current threats to the quality and quantity of water resources, this chapter will specifically address the demands, impacts and management strategies related to:

- population growth, urbanization and domestic wastewater
- industrial uses of water
- agricultural uses of water and irrigation expansion
- hydroelectric development.

The following points provide context to some of the threats listed in Table 2.1. Contamination of surface water and groundwater by the use of pesticides occurs over virtually the entire Paraná River Basin. The current data on pesticide contamination and overall water quality in surface waters in the basin is reviewed in Chapter 7 in this book. Water quality has a major impact on public health and, more broadly, on the quality of life of human populations. Water of adequate quality is also essential for the balance and functioning of ecosystems. The state of conservation of the sources and the ecosystems within the basin are determinants of the quality of the water (ANA, 2017a). In Brazil, the Upper Paraná River basin has been transformed into a sequence of reservoirs for the production of electricity, with exploitation of the high hydropower potential of the Paraná, Paranaíba, Grande, Paranapanema, Iguaçu and Tiete rivers.

The concept of water availability is associated with the provision of ecological services in the region and water uses for both consumptive and non-consumptive purposes. Given the physical and climatic range and variability within the PRB, there are all types of activities and uses of the aquatic resources, including meteorological water for agricultural activity during drought conditions, surface water to satisfy distinct consumptive uses (e.g. consumption by humans and livestock, industrial, irrigation, among others) and non-consumptive uses (e.g. hydropower generation, navigation, recreation), as well as groundwater that satisfies different types of consumptive uses, particularly for irrigation. Meteorological water availability is governed by the hydro-meteorological balance between precipitation and evapotranspiration. Surface water availability is governed by river water levels, distributed across the sub-basins. However, intensive land use leads to changes in hydrological variability of river channels and recharge of aquifers. Therefore, in this chapter, we focus on the threats associated with water quality degradation and water use conflicts, including the environmental impacts from expansion of agriculture, the lack of proper sanitation in urban areas and industrial discharges.

Population growth, urbanization and domestic wastewater

In the PRB, there are at least 20 cities with populations of more than 500,000 inhabitants, including the capital cities of Brasilia (Brazil) and Asunción (Paraguay). One of the largest metropolitan areas and industrial concentrations in the world is the city of São Paulo in Brazil, which houses more than 20 million people and is located on one of the tributaries of the Paraná River. We estimated the population within the PRB using census data gathered between 2010 and 2012, depending on the country in question (CIC, 2017a). The results indicate a total population of more than 108 million inhabitants (Table 2.2). With 42% of the total population among the four countries within the basin, Argentina and Brazil are the countries with the largest populations. Nevertheless, the entire population of Paraguay is located within the basin.

Based on changes in populations within regions of the basin recorded in censuses conducted in the four countries sharing the basin in 2000–2004 and in 2010–2012, a mean population growth rate of 10% was estimated in the basin population (Table 2.2). Growth was highest in Paraguay and Bolivia, with rates over 20% between the last two demographic censuses conducted in these countries (Table 2.2).

The demand for potable water within these urban areas is met by the large rivers of the basin, small surface water sources close to the cities, or groundwater. With the growth of cities, the available water sources, and particularly aquifers, are often overexploited or contaminated, with subsequent risks to the health of the population. The cities that present the main problems are those that have a large population and are located in the headwaters of rivers, such as São Paulo, Curitiba, Brasilia, Goiânia and Campo Grande, among others (CIC, 2017a).

Table 2.2 Populations (in millions of inhabitants) in countries that share the PRB and estimates of the populations in these countries living within the river basin. Populations were estimated from census data collected from 2010–2012. Population growth rates within these regions were estimated by comparing census data collected in 2000–2004 and 2010–2012. Estimates are based on census data from WWAP – UNESCO (2007) and CIC (2017a)

Country	*Country Population (2010–2012)*	*Estimated Population from Each Country in PRB (2010–2012)*	*Proportion (%) of total country population living in the basin*	*Estimated growth of the country population living in the basin (%)*
Argentina	41,775	29,031	69	10
Bolivia	10,598	2,064	19	20
Brazil	201,497	70,527	35	8
Paraguay	6,888	6,673	97	29
Total	**260,758**	**108,295**		**10**

Considering the data generated from a study conducted in 2013 by the National Water Agency (ANA) of Brazil regarding the water demand/availability ratio (i.e. quantitative water balance) in the Upper Paraná basin, 64% of the rivers in the Upper Paraná basin were classified as 'excellent' or 'adequate', while 36% were classified as troubling, critical, or very critical (ANA, 2015). Although the overall situation is currently adequate in terms of the quantitative water balance, the Upper Paraná basin is characterized by high population densities, with high water demands for urban and industrial use, consequently causing some of the tributaries to be in critical situations, especially those of the smaller rivers, Meia Ponte, Sapucaí, Turvo, Pardo and Mogi-Guaçu, Piracicaba and Tietê (ANA, 2015).

The lack of adequate sewage treatment is a problem that affects all countries within the basin, since only 10% of the effluents are treated (WWAP – UNESCO, 2007). In the Middle Reach of the Paraná, the impact of sewage discharges becomes evident mainly near large cities, such as Rosario. The National Water Institute (INA) in Argentina observed a high concentration of toxic blue green algae (i.e. cyanophyta) in 76% of the water samples taken from the middle reach of the Paraná River (WWAP – UNESCO, 2007). These blue-green algal species develop as a result of eutrophication of waters contaminated with sewage and industrial waste and the cyanotoxins that they produce cannot be eliminated by conventional drinking water treatment systems. The toxins produced by these algae act on the human liver and their effects are cumulative (WWAP – UNESCO, 2007).

The highest levels of sewage treatment occur in the Upper Paraná hydrographic region in Brazil, where there are the largest urban populations and the greatest number of municipalities, including six large urban agglomerates. In this region, there is removal of organic loads of greater than 60% in more than 500 municipalities, serving a combined population of over 22 million people. Even so, these discharges contribute the highest residual organic load to the Upper Paraná basin of greater than 1,487 tonnes of cBOD day^{-1} (ANA, 2017b,c).

The Tietê River sub-basin of the Upper Paraná has one of the largest resident urban populations, with roughly 29 million inhabitants. The receiving waters that fail to comply with the water quality criteria set by the Water Resources Analysis Unit (WRAU) of Brazil, which comprise 17.8% of the water bodies assessed, are located mainly in the vicinity of the Metropolitan Regions of São Paulo and Campinas, where the flow of the receiving waters is insufficient to assimilate the organic load generated from the large population (CIC, 2017b; ANA, 2017c). Overall, the hydrographic region of the Upper Paraná River stands out as the one with the highest number of municipalities and the largest population, which contributes to a low rate of dilution. That is, this region has the greatest need for advanced sewage treatment with removal efficiencies for cBOD of over 80%, or complementary solutions to improve the quality of the water resources.

In general, according to the ANA (2017a), groundwater in the Upper Paraná is of good quality and has physicochemical and bacteriological properties suited to a variety of uses, including the production of bottled mineral waters. The manner in which geological substrates store water directly influences the integrity of the groundwater. Human activities, however, can significantly compromise the quality of some aquifers. The main contaminants are nitrates, organic chemicals, and in particular, gasoline and chlorinated solvents, heavy metals, viruses and pathogenic bacteria (ANA, 2017c). The occurrence of high levels of nitrate is attributed to the lack of treatment of sewage in urban areas, and to the application of nitrogen fertilizers in rural areas. Nitrates are highly mobile in groundwater and can contaminate large areas. Agrochemicals can also contaminate groundwater, especially when there is intense agricultural activity in the recharge areas of aquifers.

Industrial uses of water

In the PRB, industries are generally associated with the processing of agricultural products. In Argentina, the major industrial districts are located in the provinces of Buenos Aires and Santa Fe. This is a significant sector, especially related to the processing of agricultural and livestock raw materials, such as tanneries, and the production of foods, beverages and milk. In Bolivia, a large part of the industry is devoted to the processing of agricultural and organic materials into end products (Genta et al., 2004; Mattos and Crespo, 2000, as cited in WWAP – UNESCO, 2007). In the part of the PRB located in Paraguay, industries are also primarily related to food and agricultural production (Dias Coelho, 2004; cited in WWAP – UNESCO, 2007). In Brazil, the largest concentration of industries lies in the southeastern and southern regions of the country. This industrial activity is highly diversified (e.g. computers and electronics, food and agricultural processing, textiles) and is mostly located in the PRB. In Brazil, industry accounts for 25% of the total water demand in the basin. The major industrial densities are located in the Tietê River sub-basin (68.5%) and especially in the São Paulo Metro Area (Dias Coelho, 2004; cited in WWAP – UNESCO, 2007). Untreated or poorly treated industrial effluents have polluted the Upper Tietê River in the vicinity of São Paulo, and also the Iguazu River in the vicinity of Curitiba (Tucci, 2004; WWAP – UNESCO, 2007).

Some of the farming-related industries are responsible for point-source discharges that reduce water quality. These agro-industries include ethanol production, and the food industries related to the production of poultry and hogs. The former industry has expanded as a result of the alcohol-for-fuel program in Brazil. Although the size of the ethanol industry in Brazil has declined over the past few years, it is still having a significant impact on freshwater quality, particularly in the state of São Paulo (Tucci, 2004). Food industries, with a strong presence in the western region of the Brazilian state of Santa Catarina, have also expanded into the states of Paraná, Mato Grosso do Sul and Goiás (Tucci, 2004; WWAP – UNESCO, 2007).

In terms of heavy metals, the contributions are mainly a consequence of industrial and mining activity. For example, contamination due to the presence of heavy metals in the Pilcomayo and Bermejo rivers originates from the extensive mining activity in the headwaters of their respective basins in Bolivian territory (WWAP – UNESCO, 2007). It is notable that mining activity in this region has taken place since pre-colonial times. Furthermore, given the large land area of the PRB and the various anthropogenic activities and their degree of development, each sub-basin presents its own challenges in terms of threats to water quality.

In Brazil, there has been a significant growth in industrial water demand in recent years, corresponding with the country's rapidly expanding economy. Water withdrawals of 207.1 $m^3 s^{-1}$ were estimated in 2013, which was the peak year over the period analyzed and 70% higher than water withdrawals in 2002. The decline in Brazilian industrial activity in 2014 and 2015 was reflected in a reduction in water withdrawals. The total withdrawal for the year 2015 was 192.4 $m^3 s^{-1}$, which was a 7.1% decrease in relation to 2013. The reduction in river flow was estimated at 104.92 $m^3 s^{-1}$, equivalent to approximately 55% of the total withdrawn (ANA, 2017b).

Agricultural use of water and irrigation expansion

Agriculture is the economic activity in the PRB that generates the greatest impacts on land use. The main crops correspond to annual cycles for production of soybeans, wheat, maize and rice. Rice is produced with irrigation from flooding and is one of the major consumers of water in the basin. The main water demands correspond to regions with flood irrigation, which are located in the Tebicuary River sub-basin in Paraguay and in southern Brazil. These regions have a large number of small reservoirs to retain water for irrigation, which represents a potential conflict with the water supply for cities in drought years. In the Center-West region of Brazil, there is a great expansion of irrigation agriculture, with sowing and harvesting occurring up to four times a year due to the great availability of water, resulting in high economic profitability. In the same region, there are conflicts over water demands associated with the expansion of sugar cane crops for ethanol production (CIC, 2017a).

Water demand, extraction and use are extremely high for agriculture, and especially for irrigation activities, as compared to other consumptive uses. The proportional participation of agriculture in water demand varies between 50% and 80%, depending on the country and the region. These percentages decrease when considering the participation of water demand and extraction for irrigation purposes relative to the total renewable water resources available in each country. Table 2.3 summarizes the total water resources and water withdrawals in each country that shares the PRB. However, it is important to note that these are national data for water withdrawals and are not exclusive to the PRB.

In the Upper Paraná in Brazil, the states in the South and Southeast rely on a dynamic, modern and diversified agricultural production of grains (soybeans, corn and wheat), as well as coffee, sugarcane, cotton and fruit farming. The determining factor governing the water resource problems observed in the rural areas of this region is the limits to agricultural expansion. Impacts are high from farms on the edge of the watercourses where there is complete removal of the riparian vegetative cover, including forested areas. There is a high level of mechanization, intensive use of agrochemicals (pesticides and fertilizers) and successive seasonal crops, with no consideration regarding environmental impacts, which include annual losses of up to 15 tonnes ha^{-1} of the topsoil, with subsequent sedimentation of watercourses (GeoBrasil, 2007).

Gonçalves et al. (2011) studied river channels in the Pontal do Paranapanema region of Brazil, which is located in the Upper Paraná basin, and observed, in areas with sugarcane plantations without appropriate soil management and intensive use of pesticides, major environmental alterations such as silting of rivers, elevation of electrical conductivity, turbidity and suspended solids in water bodies. Fernandes et al. (2018) studied limnological variables in relation to the expansion of sugarcane in a sub-basin of the Upper Paraná basin and indicated that, in years with lower production of sugarcane, there were higher levels of dissolved nutrients in the waters. Conversely, in years when the areas with sugarcane production were higher, there were higher concentrations of suspended materials in the surface waters.

Studies by Pignati et al. (2017) and Bombardi (2017) indicated that the predominant crops that use pesticides in the region are soybeans, corn and sugarcane, which account for 76% of the planted area in Brazil as of 2015. The areas with the most intensive pesticide use include

Table 2.3 National data on irrigation water use and water withdrawals for agriculture relative to total renewable water resources for the four countries that share the Paraná River Basin

Country	*Total renewable water resources (km^3)*	*Irrigation water required (km^3)*	*Water requirement ratio (%)*	*Water withdrawal for agriculture (km^3)*	*Water withdrawal for agriculture (%)*
Brazil	8,223	6.2	17	36.6	0.4
Argentina	814	3.4	16	21.5	2.6
Bolivia	622	0.3	23	1.2	0.2
Paraguay	336	0.1	23	0.4	0.1

Source: Data from WWAP – UNESCO (2007).

states located within the Upper Paraná. Roughly 899 million liters of pesticides were sprayed on these crops, and the largest quantities were used in the states of Mato Grosso, Paraná and Rio Grande do Sul, in order of use, respectively. The study by Pignati et al. (2017) also indicated that health indicators, such as acute intoxications, incidence of fetal malformation, and infant and youth cancer mortality showed a positive correlation with regional consumption of pesticides. Bombardi (2017) showed that 25% of Brazilian municipalities have their waters contaminated by agrochemicals. Data generated by SisÁgua (i.e. the database maintained by the Ministry of Health in Brazil) indicated that pesticides were detected in the public water supply in over 800 municipalities located in the states of São Paulo and Paraná. This agency monitored 27 agrochemicals and observed an increase in the quantity of substances detected over the period between 2014 and 2016 (Ministério Da Saúde, 2016).

Some studies in the Upper Paraná have focused on the variability in river flows. According to Rocha and Araújo (2016) and Rocha and Santos (2018), two components contribute to the interannual changes in the hydrological regime of the rivers: (1) rainfall variability (climatic), as a natural process over the time scale evaluated; (2) anthropogenic interferences, including altered land cover related to deforestation and implementation of several cycles of crops and livestock over time. These man-made influences can qualitatively and quantitatively modify the hydrological regime at the rainfall/river flow interface. These authors suggested that an increase in runoff during certain hydrological periods can intensify the laminar and linear erosive processes, which can be a large problem for areas with inadequate land management. These impacts in runoff regimen drive to river channel more sediment, nutrient and pesticides and become a threat for water resources.

Extensive and intensive farming activities should prioritize better management practices for soil and water management and conservation, such as no-till farming and contour cultivation, erosion containment barriers, rural road rehabilitation, maintenance or restoration of riparian forests, as well as controls for pesticide use and disposal, with specific emphasis on the activities for agri-business (GeoBrasil, 2007). In parallel, the consolidation of 'biodiversity corridors' that join together riparian forests, conservation areas and native forests, should be encouraged. In terms of groundwater, knowledge of their dynamics, including recharge areas, contamination risks, interfaces with land use management and sustainable exploitation potential, should be expanded and disseminated. Management tools specific to groundwater should be adopted and integrated with the institutional and financial provisions of the water resources policies (GeoBrasil, 2007).

Hydroelectric development

Hydroelectric development has been extensive in the PRB. For instance, major hydropower installations in the Paraná River include the Urubupungá, Porto Primavera, Itaipu and Yacyretá facilities. With regard to anthropogenic influences on water quality and quantity, a number of studies have demonstrated the effects of retention of nutrients and suspended sediments in the reservoirs of the hydroelectric developments within the PRB (Junk and Furch, 1993; Henry et al., 1999; Pagioro and Thomaz, 2002; Rocha, 2003). A report by Consórcio Oriconsul (2008) corroborated these findings on the impacts on water quality and also indicated that there has been a reduction in floodplain areas downstream of the reservoirs as a consequence of the artificially controlled water levels. The history of hydroelectric development in the basin and the impacts upon water quality and quantity are summarized in Chapter 4 of this book.

Approaches to water resource management

Water resource management includes a set of activities for the formulation of principles, preparation of technical and normative documents, implementation of management tools and organization of management, participatory and decentralized systems for implementation of water policies and decision making (Lanna, 1995; ANA, 2015a). According to Zuffo and Zuffo (2016), water resource management 'is an activity that aims to establish rights and duties, through administrative rules, so that the sustainable use of the resource is given in a fair and equitable way for all citizens involved, whether at the national or international level'.

As described in Chapter 8 in this book, water management in the Paraná River Basin should be understood within the broader scope of the La Plata Basin in terms of the multi-national treaties and pacts that have been negotiated between the countries of Argentina, Bolivia, Brazil and Paraguay to define the conditions of use of water and appropriation of the benefits generated. As noted by the CIC (2017b), 'Despite the extensive literature on conflicts over shared water use, cooperation has prevailed in the La Plata Basin, which is revealed by the number of international agreements signed and by the supranational bodies established to carry out the cooperation processes'. Villar (2015) also notes that 'the La Plata Basin reports a series of cooperation initiatives and has a range of joint institutions for the use of transboundary water resources'. The cooperation in water management in the PRB is also emphasized by Venturi (2015), who noted that the 'La Plata Treaty, which was initially conceived in the midst of geopolitical disputes, is one of understanding and cooperation', and that the 'history of sharing has another chapter of cooperation in the La Plata Basin'.

Therefore, aspects of shared water management in the entirety of the La Plata Basin and specifically in the countries covered by the PRB, are addressed within the context of the treaties and the international organizations and national and regional systems that have been created, such as the River Basin Committees. Social participation in decentralized and democratic decision-making is an important component of effective water management strategies. The sections below provide a review of the approaches that have been introduced to manage water resources within the region.

In relation to the legal and institutional aspects, the creation of the Coordinating Intergovernmental Committee of the Countries of the La Plata Basin (CIC) in 1967 and the La Plata Basin Treaty (TCP) between Argentina, Bolivia, Brazil, Paraguay and Uruguay in 1969 were key developments aimed at 'combining efforts with the goal of promoting the harmonious development and physical integration of the basin and its areas of direct and ponderable influence' (CIC, 2017b). As described by the CIC (2017a), under the Treaty, the countries agreed to promote, in the context of the basin, the identification of areas of common interest and the execution of studies, programs and works, as well as the formulation of operational understandings and legal instruments that they deem necessary and that aim at:

- Facilitation and assistance in navigation;
- Rational use of water, especially through the regulation of watercourses and their multiple and equitable use;
- Preservation and promotion of animal and plant life;
- Improvement of the terrestrial, railroad, fluvial, aerial, electrical and telecommunications interconnections;
- Regional complementarity through the promotion and creation of industries of interest for the development of the basin;
- Economic complementation of the bordering area;
- Mutual cooperation in education, health and disease control;

- Promotion of other projects of common interest, and in particular, those related to the inventory, evaluation and use of natural resources in the region;
- The integrated knowledge of the La Plata Basin.

For Brazil (2013), the La Plata Basin Treaty (PBT) 'has a clear framework nature, in accordance with other specific agreements that would be signed' and 'despite theoretical differences and conflicts, the PBT has consolidated the practice of negotiation and cooperation for the use of water resources in the region, leaving the rules, principles and institutions of an international regime'. Villar (2015) emphasizes that the La Plata Basin Treaty set the precedent for the signing of several treaties, listing 27 regional treaties between the countries of the La Plata Basin signed between 1969 and 2010).

In order to fulfill the goals of the La Plata Basin Treaty, several governing bodies were created that together compose the La Plata Basin System. Villar et al. (2018) emphasized that 'the institutional arrangements for promoting water cooperation in the La Plata Basin are fairly well developed, but not flawless' and synthesize these cooperative institutional arrangements for water management, as well as showing that 'there are 14 organizations with legal personality under international law acting on the La Plata Basin territory. Despite those organizations serving different purposes, cases of overlapping competences can be identified.' Figure 2.9 shows the areas under the jurisdiction of these organizations.

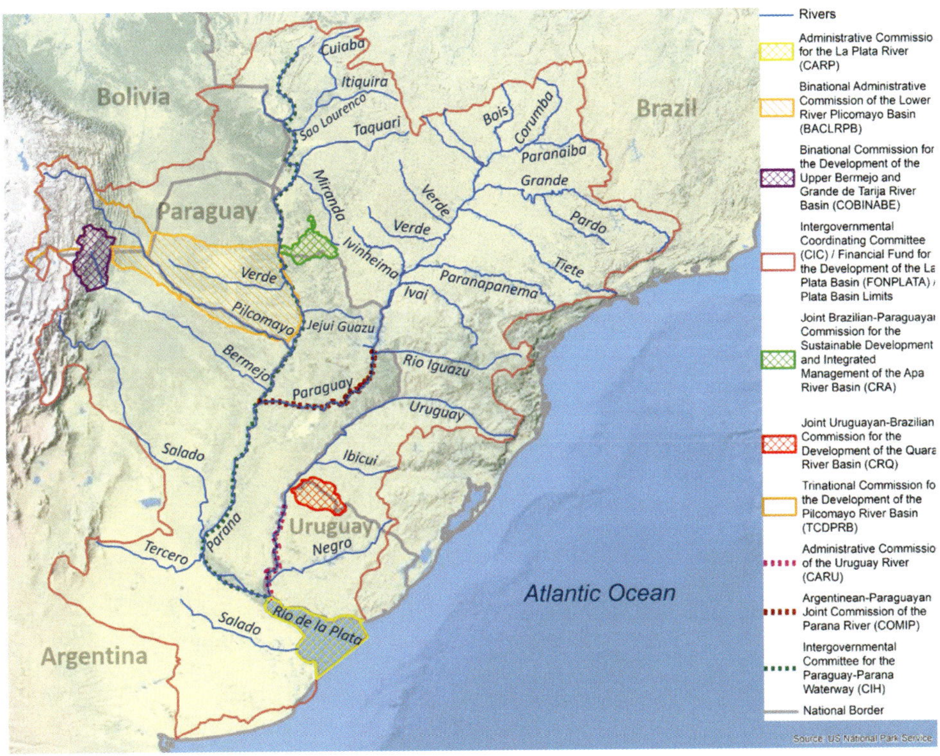

Figure 2.9 International organizations for the management of shared water resources in the Paraná River Basin.

Source: Modified from Villar et al. (2018). Erin Hayward prepared the map using ArcMap geospatial data.

In relation to the CIC, according to the Treaty, 'it has been consolidated as the body that promotes its objectives' and 'has focused on areas of common interest of the five countries, facilitating studies, programs and infrastructure works in the fields of hydrology, natural resources, transportation and navigation, soils, and energy' (CIC, 2017d, 13 p.).

Among the technical studies promoted by the CIC, in which the limits of the PRB are contemplated, one highlight is the Cross-Border Diagnostic Analysis of the La Plata Basin (CIC, 2017a), with the identification of Cross-Border Critical Issues (TCTs), their causes (as presented in the previous item in this chapter) and recommendations for management (Table 2.4) leading to combat many of the threats to water resources in the PRB.

Table 2.4 Cross-Border Critical Issues (CCIs) in the Paraná River Basin and recommendations for management

CCI	*Recommendations*
Extreme hydrological events (floods and droughts)	to consolidate, expand, and improve coordination between the various systems of monitoring, information, climate prediction, and early warning; to improve urban and territorial planning to increase resilience and reduce vulnerability to extreme events; to promote the development of regional policies and to strengthen the legal framework for the prevention and management of such events and exchange experiences with research, awareness, and environmental education programs, among other issues.
Water quality degradation	to seek funding sources for the construction and operation of domestic and industrial wastewater treatment plants, to promote the implementation of sustainable agricultural practices and the rational use of agrochemicals, and to develop programs to train environmental managers.
Sedimentation of waterways and bodies of water	to promote the development and harmonization of standards for the protection and use of natural resources; to develop land use planning and agro-ecological zoning; to strengthen institutional capacities for land-use management; to implement land reclamation and erosion control programs in priority areas; and to develop training and extension programs in soil management and conservation techniques.
Alteration and loss of biodiversity	to establish cooperation mechanisms among countries in the field of biodiversity conservation; to develop river and coastal ecological corridors and other forms of participatory conservation; to encourage the development of transboundary protected areas; and to promote the adoption of regional minimum budgets for biodiversity conservation.
Unsustainable use of fishery resources	the promotion of integrated policies, standards, and compatible criteria for the protection and sustainable use of fishery resources throughout the basin; the strengthening of management and control tools and mechanisms; the execution of vulnerability studies of priority riverine habitats, and the implementation of awareness-building and training programs on sustainable production techniques.
Unsustainable use of aquifers in critical areas	to develop integrated and participatory management tools; to carry out vulnerability studies to identify areas of risk at the regional and local levels; and to develop regional inventories and databases and encourage greater social participation.

Water use conflicts and environmental impact of irrigating crops	to promote agreements and the development of common legal frameworks for water use management; to strengthen the management capacity and institutional coordination of relevant agencies in the five countries; to generate information and facilitate public access to data for managing supply and demand; and to establish strategies for communication, dissemination of information, and public awareness about management.
Lack of disaster contingency plans	are to establish common safety standards and criteria, considering the incidence of climate variability and change; to develop and adopt national and transnational safety and operational standards for emergencies; to develop or update contingency plans and programs in the event of dam rupture; and to develop awareness-raising measures amongst citizens on prevention and risk reduction.
Poor state of waterborne health parameters and deterioration of sanitation	strengthening research and the generation and dissemination of data on waterborne diseases; promoting policies and programs for the treatment of solid waste, industrial waste, and the management of agrochemicals; strengthening the capacity of local managers and the articulation and institutional coordination of institutions in the water and sanitation sectors of the countries; and to promote education and public awareness programs on environmental hygiene and health.
Navigational limitations	to harmonize regional policies for river transport; to adapt the legal and institutional framework for river navigation; to develop transboundary plans for the maintenance and dredging of inland waterways; and to promote an integrated transport system.
Development of hydropower potential	to make agreements for energy integration between the countries of the basin; to integrate hydro-meteorological monitoring networks from hydraulic use to other information systems; and to take actions to take advantage of the regional interconnected communications system in order to improve the transmission of information for early warning hydrological systems.

Source: CIC (2017a, 2017e).

The recommendations formulated on the basis of the CCIs and grouped under different aspects were subsequently consolidated into six (6) Strategic Areas (CIC, 2017c):

- Information management;
- Planning, management and sustainable handling of water resources;
- Environmental protection/rehabilitation;
- Education, communication and public participation;
- Research and technological development; and
- Institutional strengthening.

Cross-Border Critical Issues and Strategic Actions should underpin the performance of the La Plata Basin System and national water management systems. It is important to also consider the impacts and threats in the waters resulting from the use of agrochemicals in agriculture and the exploitation of shale gas, along with its control and/or restriction.

In the context of the work of the CIC, the La Plata Basin Program has the general objective of:

> Strengthening cross-border cooperation between the governments of Argentina, Bolivia, Brazil, Paraguay and Uruguay in order to ensure the management of shared

> water resources in an integrated and sustainable way of the basin, in the context of climate variability and change, capitalizing opportunities for development.
>
> (CIC, 2017d)

Ribeiro (2017) analyzed the Treaty for Hydropower Utilization of Water Resources of the Paraná River of Salto Grande de Sete Quedas or Salto de Guaíra, 1973, noting that:

> It established the creation of a binational entity, Itaipu, responsible for the 'design, construction and operation of a power plant to generate electricity', and that 'even before the signing of the agreement in Brasília, the two countries had already cooperated to carry out studies on the hydropower potential of the Paraná River'. That is, the Itaipu Treaty, as it was known, reaffirmed earlier cooperation between the parties.

He emphasized that the 'terms agreed in the Itaipu Treaty, defined by dictatorial governments in both countries, made clear the asymmetry between the parties. The document led to a series of Paraguayan inquiries over decades, culminating in 2008 with the revision of the terms of the Treaty' (Ribeiro, 2017), with Brazil partially meeting Paraguay's demands to revise the Agreement during the Nicanor administration, which later turned to energy sovereignty in the words of President Lugo, with the change (increase) of the tariff. This agreement will be reviewed this year, which may lead to a new crisis between countries.

In relation to internal water management in the four countries of the PRB, in addition to environmental management policies and systems with interfaces in water resources, Argentina, Bolivia, Brazil and Paraguay have policies and systems for the management of water resources that incorporate principles and objectives that are in line with definitions of global organizations, including the adoption of the systemic model of participatory integration in water management. According to Setti et al. (2000), this model is:

> characterized by the creation of a systemic structure, in the form of an institutional management matrix, responsible for the execution of specific managerial functions, and by the adoption of three instruments: INSTRUMENT 1: Strategic planning by hydrographic basin (. . .); INSTRUMENT 2: Decision making through multilateral and decentralized releases (. . .); and INSTRUMENT 3: Establishment of legal and financial instruments.
>
> (Setti et al., 2000)

The description of the community-based approach focuses primarily on the activities of River Basin Committees in Brazil. These decentralized organizations allow integration and social participation in water management system, as well as with other organizations involved in water governance (e.g. communities, NGOs, associations, etc.). The River Basin Committees stand out, in the Brazilian case, as decentralized organizations that allow integration and social participation in the water management system. Water governance, according to Villar (2015) 'gained prominence in the context of the water crisis, but despite its profusion in literature, there is no consensus on its concept'. Similarly, Ribeiro (2016) notes that 'governance has been used mainly as an umbrella concept, and there is no consensus definition'. Mauro et al. (2017) address governance in the context of the National Water Resources Management System for Brazil (SINGREH).

In the PRB, the River Basin Committees are expected to play a key role in the process of democratic water management. In relation to social participation in water decisions, the water crises that have reached the Southeast region of Brazil, with impacts on the water supply of the São Paulo Metropolitan Area, on the Upper Tietê River, a tributary of the Paraná River, which occurred in 2014 and 2015, have raised concerns about how they are managed, with centralization of information and decisions. These crises are strongly manifested by reductions in precipitation but derive from several other reasons, as noted by Buckeridge and Ribeiro (2018):

> The causes of water crises clearly have a systemic character. That is, one should not interpret the occurrence of such a phenomenon based on unique causes, but rather on multiple factors that interact and lead to lack of water for the population. One of the main causes is the increase in consumption observed in the Region corresponding to Greater São Paulo during the 20th century.

In 2004, the renewal of the concession for Basic Sanitation Company of the State of São Paulo, Brazil (SABESP) to maintain the water catchment in the Piracicaba River basin, to enable the Cantareira System to supply a significant part of the São Paulo Metro Area resulted in the intense participation of the CBH for Piracicaba, Capivari and Jundiaí (PCJ). For the first time, a grant from the State of São Paulo was provided, with the inclusion of conditioning measures, duly agreed with the participation of the ANA, the São Paulo State Secretariat, SABESP, the Department of Water and Electricity for the State of São Paulo, Brazil (DAEE) and the CBH for the PCJ. In this case, the conditions required of the user, SABESP, were not met in their entirety. If these conditions had been met, certainly the water crises that followed would not have occurred or would have had less negative impacts than those recorded (Di Mauro, 2017). Thus, it is understood that politicians who benefit from the powerful socioeconomic sectors are largely responsible for the failures during water crises, and this case history demonstrates the weaknesses of SINGREH in Brazil. There is a need to change water management procedures and implement programs and actions provided in water resources plans in order to avoid water crises. Robust strategies to avoid these crises were proposed in the Paranapanema Integrated Water Resources Plan developed by the regional Water Resource Management Unit (UGRH), namely: water availability, regulation of water demands, and regulation of pollutant loads (ANA, 2016).

Another key issue for water resource management and social participation is the safety of dams, especially after the rupture of mining tailings dams in the municipalities of Mariana and Brumadinho in the state of Minas Gerais, leading to loss of life, destruction of infrastructure and significant environmental impacts on water bodies (Freitas et al., 2019). Thus, a number of surveys have been carried out on the impacts of dams in Brazil, including in the Paraná River basin. From the standpoint of participatory management of waters, the Cultivando Água Boa Program of the Paraná River Basin from Itaipu Binacional stands out, having been recognized as the best water management project in the world in 2015. In addition, the integrated management of the Guarani Aquifer is worth mention as a successful approach to water management.

Transboundary management of the La Plata basin

There are many treaties and institutions that have been developed to manage various aspects of water resources within the La Plata Basin. The La Plata Treaty of 1969 sought to regulate the use of waters of the La Plata Basin. Before it, the Coordinating Intergovernmental Committee

(CIC) was established in 1967, and much later, in 1991, the Mercosur Bloc established a Common Market Council that could face challenges that generate tension between the countries of the La Plata Basin.

The Itaipu Treaty, signed between Brazil and Paraguay in April 1973, laid the foundations for the construction of the Itaipu Hydropower Power Plant. Relevant aspects were also agreed upon, such as Paraguay's indebtedness to Brazil, which financed the project, and the commitment to sell the unused energy surplus to the other partner in the project. This agreement reflects the Brazilian hydro-hegemony at the time. It was an agreement signed in the Cold War era, in which countries sought to establish alliances with global powers. Brazil, like Paraguay, was governed by a military dictatorship aligned with the USA. An intervention of this magnitude in South America also represented greater control over Argentina, at the time defined as the main foreign enemy of Brazil. That is, Brazil sought to attract Paraguay to its scope of influence, isolating it from Argentina.

In fact, this strategy was not successful, since Paraguay later exercised the option to negotiate with Argentina as well. This strategy resulted in the Itaipu Corpus Tripartite Agreement, signed in October 1979, which paved the way for the construction of three hydropower plants on the Paraná River, namely, Itaipu, Corpus and Yaciretá-Apipé (Caubet, 2006). Through this agreement, Argentina began to demand compliance with the principle of prior consultation for the use of the waters of the Paraná River.

Despite the tensions that this demand generated, the Itaipu hydropower plant was built, involving Brazil and Paraguay, as well as the Yaciretá-Apipé hydropower plant, with the joint participation of Argentina and Paraguay. A third hydropower plant, Corpus, was designed, between Paraguay and Argentina, which has not yet been built. A series of tensions that developed between Brazil and Paraguay emerged at the beginning of the 21st century. Paraguay began to claim, under the government of President Nicanor Duarte, a revision of the terms of the Itaipu Treaty. Later, with President Fernando Lugo, the demands focused around two central issues: (i) revision of the price of the tariff paid by Brazil to Paraguay and (ii) energy sovereignty, whereby Paraguay claimed the freedom to sell surplus energy on the open market. The negotiations conducted by Lugo and then Brazilian President Luiz Inácio Lula da Silva resulted in improvements to the tariff paid by Brazil to Paraguay but maintained the obligation to sell the surplus to the other partner. This demand is a topic of discussion during the current negotiations to revise the Itaipu Treaty. In addition, the Paraguayan debt is expected to be settled in 2022, which may also change the country's position within the Itaipu Binacional agency (Ribeiro, 2017). It remains to be seen whether countries will continue to neglect established institutions or if they will be able to negotiate durable outflows without direct confrontation between the parties.

Conclusions

Intensification of irrigation for monocultures is a threat to the availability of water resources and to water quality as a result of agrochemicals in runoff and sedimentation resulting from poor management strategies. Agrochemicals include fertilizers and pesticides. The combined effects of multiple pesticides and transformation products are often not known, leading to uncertainty regarding sustainable management of water resources. Reductions in the quantities of water that are available can lead to possible conflicts in areas of intensive irrigation in the PRB. These threats could be exacerbated by climate change and changes in hydrological cycles.

Other threats to water quality encountered in the PRB are derived from the introduction of substances that degrade water quality through the release of untreated or poorly treated sewage and industrial effluents into surface waters. It should be noted that many contaminants of emerging concern are not monitored by environmental agencies, although they are extremely biologically active, such as hormones and some antibiotics. Another noteworthy threat is from the disposal of solid waste, which promotes contamination of surface and groundwater.

The history of water management strategies in the PRB has provided experiences that point to the value of using an integrated management approach for shared waters, involving the work of several international, national and regional organizations. There are major challenges to be overcome, such as addressing the Cross-Border Critical Issues that have been identified and other existing problems and tensions in order to guarantee water sustainability and democratic management of shared water.

The various treaties and agreements are important instruments for solving domestic and transboundary problems, as well as for finding solutions to threats to water resources. The formation of river basin committees is a useful model for addressing threats to water resources, provided that their composition maintains participation by both government and the public. This democratization is of fundamental importance for conflict resolution.

References

Amore, L. 2011. The Guarani aquifer: From knowledge to water management. *Water Resources Development* 27:463–476.

ANA. 2015. *Conjuntura dos recursos hídricos no Brasil: Regiões hidrográficas brasileiras – Edição Especial.* Agência Nacional de Águas, Brasília, Brazil.

ANA. 2016. *Plano integrado de recursos hídricos da UGRH Paranapanema.* Paranapanema River Basin Committee, Agencia Nacional de Aguas, Brasilia, Brazil.

ANA. 2017a. *Conjuntura dos recursos hídricos no Brasil*, Relatório pleno, Agência Nacional de Águas, Brasília, Brazil.

ANA. 2017b. *Água na indústria: Uso e coeficientes técnico.* Agência Nacional de Águas, Brasília, Brazil.

ANA. 2017c. *Atlas Esgotos: Despoluição de bacias hidrográficas, Secretaria Nacional de Saneamento Ambiental.* Agência Nacional de Águas, Brasília, Brazil.

Barros, V.R., Boninsegna, J.A., Camilloni, I.A., Chidiak, M., Magrín, G.O., Rusticucci, M. 2015. Climate change in Argentina: Trends, projections, impacts and adaptation. *Wiley Interdisciplinary Reviews: Climate Change* 6:151–169.

Bombardi, L.M. 2017. *Geografia do Uso de Agrotóxicos no Brasil e Conexões com a União Europeia/Larissa Mies Bombardi*, FFLCH – USP, São Paulo, 296 p. Retrieved from: http://conexaoagua.mpf.mp.br/arquivos/agrotoxicos/05-larissa-bombardi-atlas-agrotoxico-2017.pdf Accessed in May, 2019.

Brião, V.B., Magoga, J., Hemkemeir, M., Brião, E.B., Girardelli, L., Sbeghen, L., Favaretto, D.P.C. 2014. Reverse osmosis for desalination of water from the Guarani Aquifer System to produce drinking water in southern Brazil. *Desalination* 344:4012–4411.

Buckeridge, M., Ribeiro, W.C. 2018. *Livro branco da água. A crise hídrica na Região Metropolitana de São Paulo em 2013-2015: Origens, impactos e soluções*, 1st ed. Instituto de Estudos Avançados, São Paulo, vol. 1, 175 p.

Caubet, C. 2006. *A água doce nas relações internacionais.* Manole, São Paulo, Brazil.

CIC. 2017a. *Análise diagnóstico transfronteiriço da Bacia do Prata*, 1st revised ed. Comité Intergubernamental Coordinador de los Países de la Cuenca del Plata, Organization of American States (OAS), Buenos Aires, Argentina.

CIC. 2017b. *Marco institucional y legal para la gestión integrada de los recursos hídricos en la cuenca del Plata*, 1st ed. Comité Intergubernamental Coordinador de los Países de la Cuenca del Plata, Organization of American States (OAS), Ciudad Autónoma de Buenos Aires.

CIC. 2017c. *Programa de ações estratégicas da Bacia do Prata*, 1st special ed. Comité Intergubernamental Coordinador de los Países de la Cuenca del Plata, Organization of American States (OAS), Ciudad Autónoma de Buenos Aires.

CIC. 2017d. *Análise Diagnóstico Transfronteiriço e Programa de Ações Estratégicas da Bacia do Prata: Resumo executivo*, 1st ed. Comité Intergubernamental Coordinador de los Países de la Cuenca del Plata, Organization of American States (OAS), Ciudad Autónoma de Buenos Aires.

CIC. 2017e. *Transboundary Diagnostic Analysis for the La Plata River Basin (TDA)*, 1a edición especial. Comité Intergubernamental Coordinador de los Países de la Cuenca del Plata, Organization of American States (OAS), Ciudad Autónoma de Buenos Aires.

Collischonn, W., Bravo, J.M., Pontes, P.R.M. 2016. Paraná River Basin. In: Singh, V.P., *Handbook of Applied Hydrology*, 2nd edition, McGraw-Hill Education, New York, vol. 1, p. 1440.

Consórcio Oriconsul. 2008. *Diagnóstico da região da Bacia do rio Uruguai-Relatório Final do Componente 1.* Florianópolis/SC, Porto Alegre/RS, Consórcio Oriconsul – Ecoplan – Skill, São Paulo, Brazil, 550 p.

Di Mauro, C.A. 2017. Outorga para captação de hecursos Hídricos no sistema cantareira e vulnerabilidade da população na região Metropolitana de São Paulo. In: Chávez, E.S., Mauro, C.A., Di Moretti, E.C., (Coord.), Dias, L.S., (Colab.), *Água, Recurso Hídricos Bem Social Transformado Em Mercadoria*, ANAP, Tupa, São Paulo, pp. 32–45.

Fernandes, M.R., Araújo, R.R., Rizk, M.C., Santos, A.A. 2018. *Análise limnológica e a expansão da cana-de-açúcar em uma sub-bacia hidrográfica do rio Santo Anastácio, Piquerobi/SP*, Volume 2: III CIPEEX, Ciência para a redução das desigualdades, IX Simpósio Nacional de Ciência e Meio Ambiente, Edição, pp. 8–18.

Freitas, C.M., Barcelos, C., Asmus, C.I.R.F., Silva, M.A., Xavier, D.R. 2019. Da samarco em mariana à Vale em Brumadinho: Desastres em barragens de mineração e saúde coletiva. *Cad. Saúde Pública* 35: e00052519.

GeoBrasil. 2007. *Recursos hídricos: Resumo executivo, Ministério do Meio Ambiente, Programa das Nações Unidas para o Meio Ambiente.* Agência Nacional de Águas, Brasília, Brazil, 60 p.

Gonçalves, F., Rocha, P.C., Ferreira, C.C. 2011. Uso e ocupação da terra e suas influências em parâmetros químicos e físicos da água da bacia hidrográfica do rio Santo Anastácio, Oeste Paulista. *Proceedings of the 15th Brazilian Conference on Remote Sensing held in Curtiba*, Brazil, April 30 to May 5, 2011, pp. 1248–1255.

Henry, R., Santos, A.A.N., Camargo, Y.R. 1999. Transporte de sólidos suspensos, N e P total pelos Rios Paranapanema e Taquari e uma avaliação de sua exportação na Represa de Jurumirim (São Paulo, Brasil). In: Henry, R. (ed.), *Ecologia de reservatórios: Estrutura, função e aspectos sociais*, Botucatu: FUNDIBIO: FAPESP, pp. 153–162.

Junk, W.J., Furch, K.A. 1993. A general review of tropical South American floodplains. *Wetlands and Ecological Management* 2:231–238.

Krepper, C.M., Venturini, V. 2009. Assessing interannual water balance of La Plata river basin. *Atmósfera* 22:387–398.

Lanna, A.E.L. 1995. *Gerenciamento de bacia hidrográfica: Aspectos conceituais e metodológicos.* IBAMA, Brasília, 170 pp.

Lenhard, L.G., Anderson, S.M., Coimbra-Araújo, C.H. 2018. Energy-environmental implications of shale gas exploration in Paraná hydrological basin, Brazil. *Renewable and Sustainable Energy Reviews* 90:56–69.

Mauro, C.A., Di Mageste, J.G., Lemes, E.M. 2017. As bacias Hidrográficas como critério para o Planejamento Territorial. In: *Caminhos de Geografia – Edição especial – Gestão de Bacias Hidrográficas: Da Teoria à Praxis.* Caminhos de Geografia, 18:64, pp. 472–482.

Meis, M., Llano, M.P. 2019. Hydrostatistical study of the Paraná and Uruguay Rivers. *International Journal of River Basin Management* 17:1–12.

Ministério Da Saúde. 2014. *Monitoramento de agrotóxicos em água para consumo humano, Brasil. 2016*, Boletim Epidemiológico, Secretaria de Vigilância em Saúd, 47:12, Brazil.

OAS. 2005. *La Plata River Basin*, Water Project Series, Number 6, Office for Sustainable Development and Environment, Organization Of American States (OAS), Washington, D.C., USA.

OAS/DSD. 2019. *Sustainable Management of the Water Resources of the La Plata Basin with respect to the Effects of Climate Variability and Change*, Organization Of American States, Department Of Sustainable Development (OAS/DSD). Retrieved from: www.oas.org/dsd/WaterResources/projects/LaPlata_NEW_eng.asp.

Pagioro, T.A., Thomaz, S.M. 2002. Longitudinal patterns of sedimentation in a deep, monomictic subtropical reservoir (Itaipu, Brazil-Paraguai). *Archives of Hydrobiology* 154:515–528.

Pignati, W.A., Souza, E., Lima, F.A.N., Lara, S.S., Correa, M.L.M., Barbosa, J.R., Leão, L.H.C., Pignatti, M.G. 2017. Spatial distribution of pesticide use in Brazil: A strategy for health surveillance. *Ciência & Saúde Coletiva* 22:3281–3293.

Puig, A., Salinas, H.F.O., Borús, J.A. 2016. Relevance of the Paraná River hydrology on the fluvial water quality of the Delta Biosphere Reserve. *Environmental Science and Pollution Research* 23:11430–11447.

Ribeiro, N.B. 2016. *Governança sistêmica das águas: Proposição de um modelo analítico e sua aplicação na Bacia Lagos São João, RJ. Tese de Doutorado*, UERJ, Rio de Janeiro.

Ribeiro, W.C. 2017. Shared use of transboundary water resources in La Plata River Basin: Utopia or reality? *Ambiente E Sociedade (Campinas)* 20:257–270.

Rocha, P.C., Araújo, A.P. 2016. O regime hidrológico na unidade de gerenciamento de recursos Hídricos paranapanema: Variabilidade interanual e espacial. In: *Leonice Seolin Dias e Sandra Medina Benini, Estudos ambientais aplicados em bacias hidrográficas*, 2nd edition. ANAP, Tupa, p. 218.

Rocha, P.C., Santos, A.A. 2018. Hydrological Analysis in Water Basins. *Mercator Fortaleza* 17:e17025.

Rocha, R.R.A. 2003. *Limnologia comparativa e preditiva de rios e lagoas da planície de inundação do alto rio Paraná*. Doctoral Thesis, Universidade Estadual de Maringá, Maringá, Brazil.

Setti, A.A., Lima, J.E.F.W., de Miranda, A.G., de Pereira, I.C. 2000. *Introdução ao gerenciamento de recursos hídricos*, 2nd edition. Agência Nacional de Energia Elétrica, Superintendência de Estudos e Informações Hidrológicas, Brasília.

Tucci, C. 2004. *Visão dos Recursos Hídricos da bacia do Rio da Prata. Visão regional.*, Programa Marco para la Gestão Sustentável de los Recursos Hídricos da Bacia do Prata., Organization of American States, Washington, D.C., USA, 227 pp.

Venturi, L.A.B. 2015. *Água no Oriente Médio: O fluxo da paz*, 1st edition. Editora Sarandi, São Paulo, Brazil.

Villar, P.C. 2015. O papel do Comitê Intergovernamental Coordenador da Bacia do Prata (CIC) na cooperação internacional das águas transfronteiriças. *Proceedings of the 5th National Opening Meeting held in Belo Horizonte*, Minas, Brazil, pp. 1–12.

Villar, P.C., Ribeiro, W.C., Sant'anna, F.M. 2018. Transboundary governance in the La Plata River basin: Status and prospects. *Water International*. DOI:10.1080/02508060.2018.1490879.

WWAP – UNESCO. 2007. La Plata Basin Case Study. World Water Assessment Programme – United Nations Educational, Scientific and Cultural Organization. Retreived from: https://unesdoc.unesco.org/ark:/48223/pf0000151252 Accessed in May 7, 2019.

Zuffo, A.C., Zuffo, M.S.R., 2016. *Gerenciamento de recursos hídricos: conceituação e contextualização*. Elsevier, Rio de Janeiro/ Amsterdam, The Netherlands, p. 456.

3

Geomorphological and sedimentological characteristics

Oscar Orfeo, José Stevaux, and Daniel Parsons

Key messages

- The Upper Reach of the Paraná River runs over a mixed rock alluvial bed, with a relatively narrow alluvial plain, two alluvial terraces and several breaks in channel slope in its longitudinal profile. Although the highest elevation in the basin is 1,200 m at the headwaters of the Grande River, it is generally recognized the Upper stretch of the Paraná River begins at the confluence of the Paranaíba and Grande rivers at an altitude of 290 m above sea level and extends for 600 km to the reservoir of the Itaipu Dam.
- At the Guaira gauge station in the Upper basin, the mean discharge is 10,800 $m^3 s^{-1}$, with extreme minimum and maximum values of 2,490 $m^3 s^{-1}$ (1944) and 39,870 $m^3 s^{-1}$ (1983),

respectively. Because of the heavier rainfall in the eastern part of the basin, all the largest tributaries are situated on the eastern border, with mean discharges in these tributaries of between 860 and 3,000 m^3 sec^{-1}.

- Owing to sediment deposition in the reservoirs of the numerous dams in the Upper Paraná, there are very low loads of suspended solids in the main channel. For instance, in the period before and after the construction of the Porto Primavera Dam, the mean concentrations of suspended sediments declined from 35 mg L^{-1} before construction to 10.85 mg L^{-1} at the completion of the dam, and today the mean suspended sediment concentration is 0.3 mg L^{-1}.
- The multichannel pattern of the Upper Paraná is characterized by more than 200 islands of different morphology, size and age, plus hundreds of central and lateral sand bars. Some sections of the river present up to six multichannel stretches, although a main eastern channel is practically continuous and constitutes the primary navigation way.
- Downstream of the basaltic plateau located at the Brazil-Argentina border, the Paraná River flows through the extensive Argentine plain. Andean tectonics have modified the regional slope and drainage to form the extensive Iberá wetland.
- Downstream of the confluence with the Paraguay River, the Middle Reach of the Paraná shows pronounced spatial and temporal differences in suspended sediment concentrations, related to the sediments transported from the Andean region by the Paraguay River. The suspended sediment concentration in the Paraná River downstream of the confluence with the Paraguay River shows extremes up to 616 mg L^{-1} on the eastern margin and up to 1,271 mg L^{-1} on the western margin.
- The present multichannel pattern of the middle Paraná River is related to the abundance of the bedload discharge, which is greater than 25% of the total sediment discharge. Because of this and the regional slope, several kinds of sand bars and channels are located in the middle Paraná as a typical feature of this fluvial system.
- In the Lower Paraná, the Delta region has a surface area of around 3,500 km^2 and it is composed of islands with a central marsh or shallow lake, surrounding by meandering streams with sandy natural levees, where scroll meanders, point bars and oxbow lakes are also frequent.

Introduction

A river system consisting of a network of channels separated by small islands that are often temporary is called a braided river. Sandy, braided rivers form some of the most important alluvial channels in the world, ranging in scale from small braided rivers in upland valleys, to spectacular, large braided channels, such as the Paraná River. The alluvium of braided rivers is also economically important as repositories for water, oil, gas and minerals, and therefore the sedimentological architecture of braided rivers is of key importance for exploitation of these reserves (Best et al., 2006).

The main course of the Paraná River is normally subdivided into the following stretches: Upper, Middle, and Lower (Figure 3.1a). The Upper stretch is situated mainly in Brazil and extends from its headwaters to the confluence with the Paraguay River, near Corrientes City in Argentina. The Paraguay River is the main tributary of the Paraná River (Figure 3.1a), which increases its discharge by more than 40% and the suspended sediment discharge from 5.1×10^6 tons $year^{-1}$ to 118.7×10^6 tons $year^{-1}$ (Orfeo and Stevaux, 2002). The Middle stretch runs from that confluence with the Paraguay to Diamante City in Entre Ríos province of Argentina along more than 700 km, developing a wide alluvial plain (Figure 3.1a). At that last geographical point, the Lower stretch begins and extends to the Rio de la Plata estuary, including the Delta with a surface area of 14,000 km^2. The total length of the Paraná reach is nearly 4,000 km from the Brazilian highlands to the mouth of the basin in Argentina. The highest elevation of the river basin is located in the headwaters of the Grande River and the elevation declines steadily over the Upper Reach to the confluence of the Paraguay River and Paraná River at Corrientes (Figure 3.1b).

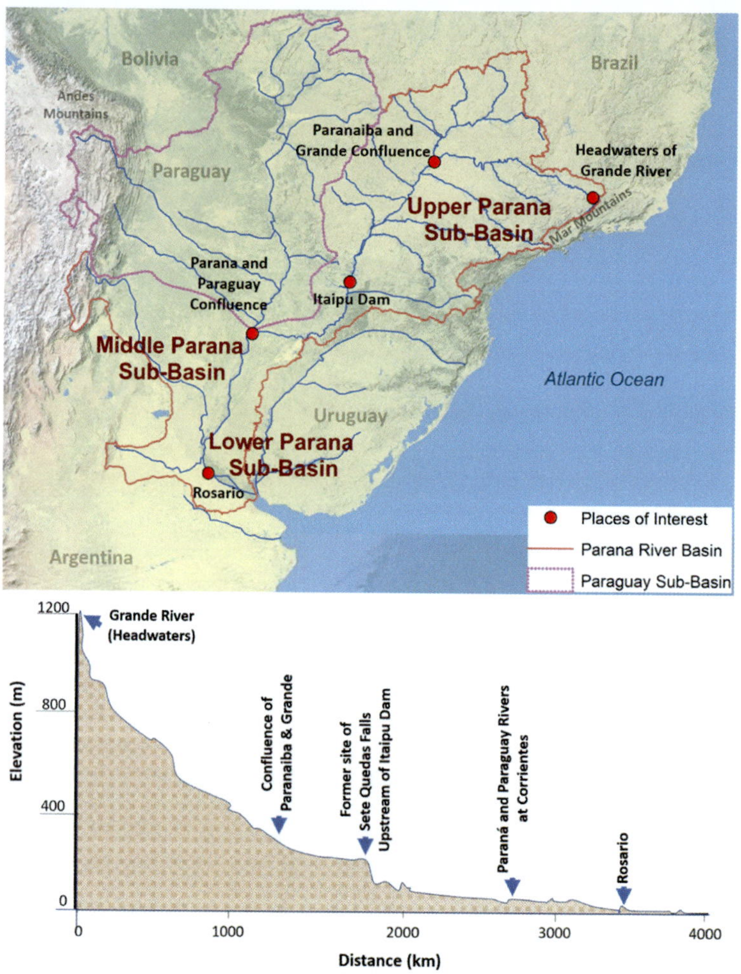

Figure 3.1 The Paraná River (Upper, Middle and Lower Reach) and the Paraguay River sub-basin (a) and the longitudinal profile from the headwaters of the Grande River to the discharge into the Rio de la Plata estuary (b). Note the vertical drop at the former site of Sete Quedas Falls, which is now the site of the Itaipu Dam, which lies between the upper and middle courses of the river. Hamid Mehmood compiled the geospatial data in ArcMap format and Erin Hayward prepared the map using those data.

Upper Paraná River Basin

Morphology and discharge

In its upper reach, the Paraná is a plateau river and runs over a mixed rock (mainly basalt and sandstone) alluvial bed, with a relatively narrow alluvial plain (in general < 4 km, with a maximum of 13 km), thin alluvial sediments (< 20 m in thickness), two alluvial terraces and several small knickpoints in its longitudinal profile. The channel slope is relatively low, varying from 8 to 3 cm km^{-1}. The Upper Paraná River (UPR) is defined as beginning at the confluence of the Paranaíba and Grande rivers at an altitude of 290 m above sea level and extending for 600 km up to the former site of Sete Quedas Falls, an impressive 200 m knickpoint, which today is covered by the reservoir of the Itaipu Dam. After that, the river runs for 211 km to cross the Brazil-Argentina border. However, the headwaters of the Grande River are located at the Mar Mountain, 1,200 m above sea level (Figure 3.1b).

At the Guaira gauge station, which is the most downstream of the UPR and has been active since 1910, the mean discharge is 10,800 m^3 s^{-1}, with extreme minimum and maximum values of 2,490 m^3 s^{-1} (1944) and 39,870 m^3 s^{-1} (1983), respectively. The UPR Basin is asymmetrical, with the area of the eastern border being larger than the western border, and annual rainfall on the eastern border (1,200–2,200 mm) being almost twice as great as the rainfall on the western border (1,200–900 mm). Therefore, all the largest tributaries of the system are situated on the eastern border, with mean discharges of between 860 and 3,000 m^3 sec^{-1} (Table 3.1).

Sedimentary load

Because of the deposition of sediments that occurs in the reservoirs of the large number of dams constructed both in the main river channel and in its primary tributaries, the Upper Paraná River Basin has a relatively low load of suspended sediments. Studies of bedload transport in the river have been conducted in nodal sections (Martins and Stevaux, 2005) and in multichannel segments (Gon, 2012). Bed sediment consists predominantly of medium to coarse quartzose sand. Some fine gravel deposits were deposited under past flow conditions and are not currently being transported (Santos et al., 2017). Bed loads are transported mainly through dune and mega ripples (Figure 3.2), although plan bed and ripples can occur under specific flow conditions. Measurements made in nodal sections indicated that dunes drift at a velocity of 1.0 to 1.9 m day^{-1}, resulting in daily transport of 2,830 t of material (Martins and Stevaux, 2005)

Table 3.1 Mean discharge and surface area of the sub-basin for the major tributaries of the Upper Paraná Basin, listed according to the size of the sub-basin.

Tributary	*Mean Discharge(m^3 s^{-1})*	*Basin Area(km^2)*
Paranaiba River	3,000	222,000
Grande River	2,100	143,000
Paranapanema River	1,400	109,000
Tiete River	602	74,100
Iguazu River	1,542	69,000
Ivaí River	860	34,000

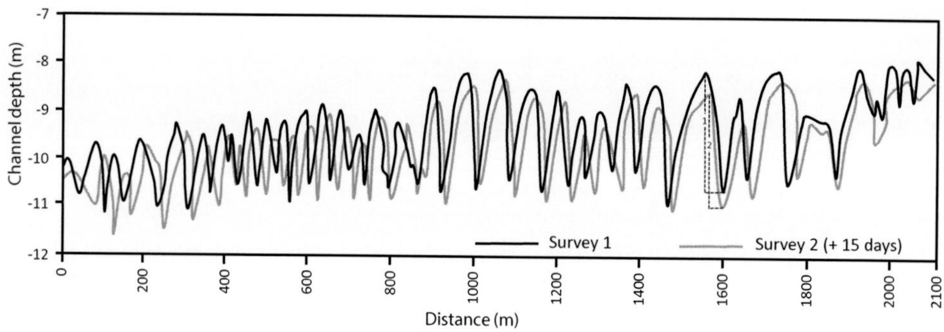

Figure 3.2 Echo-bathymetrical longitudinal profiles of channel depth (m) in the Upper Paraná River generated in two surveys conducted over a 20-day interval. Channel bedforms mainly consist of dunes of 2 m height and 100 m length. This configuration in the channel no longer exists due to the operation of the Porto Primavera Dam, as described in the text. The figure is modified from Martins and Stevaux (2005).

Channel pattern and morphology

Of the 811 km of the Upper Paraná River in Brazil, only a 235 km stretch of the river between Porto Primavera Dam and the beginning of the reservoir of the Itaipú Dam remains in its 'natural' (i.e. not dammed) condition. The Upper Reach of the Paraná is litho-structurally controlled by northeast to southwest and east to west fault systems associated with outcrops of the hard sandstone of the Caiuá Formation (Cretaceous) that extends almost continuously along the eastern border. Although the faults are not currently active, they control the rectilinear morphology of the floodplain, and associated with the lithology, produce an asymmetrical 4 to 15 km floodplain and two terraces that are present mainly on the western border and a 20 m height steep left bank (Souza et al., 2013; Fortes et al., 2005; Souza Filho and Stevaux, 2004; Stevaux, 1994).

The multichannel pattern of the Upper Paraná River is characterized by more than 200 islands of different morphology, size and age, plus hundreds of central and lateral sand bars. Some sections of the river present up to six secondary channels, although a main eastern channel is practically continuous over the entire reach. Leli (2015) described two types of islands: those generated by in-channel processes (i.e. sedimentation in the channel) and those resulted from off-channel processes by avulsion and floodplain cut-off (Figure 3.3). In general, large islands are highly permanent in the channel. From analysis of sediments on large islands, Leli (2015) estimated ages of up to 8,200 years, suggesting a relatively stable hydro-sedimentary dynamic since the early Holocene.

Avulsion islands, or islands formed by a cutoff from the floodplain, include the larger types of 50 to 400 km^2 area. The avulsion processes in the Upper Paraná River have not been well studied, but some authors attribute them as originated by neo-tectonic activity (Fortes et al., 2005). These types of islands have a concave upward morphology with natural levees of 2 to 4 m developed at the borders. The islands are normally covered by grassy and shrubby vegetation, except for treed areas on the natural levees. Many sectors of these islands are swampy and it is possible to identify some structures as relict morphologies of ancient channels, levees and islands from the original floodplain.

Islands generated by central bar sedimentation are smaller but more frequent than avulsion islands and are associated with the hydrodynamic conditions of the river. Vegetation plays an

Figure 3.3 Satellite image of a typical multichannel pattern of the Upper Paraná River formed by island of different morphology, size, origin and age (Google Earth).

important role in the establishment and evolution of these types of islands (Stevaux, 1994). Once fixed, a sand bar is progressively covered by fine sediments that are deposited annually during ordinary floods. With the continuation of the process, the surface of the bar rises by vertical aggradation of fine sediments, up to the level of the riverbank, where the process is stabilized.

Although the Upper Paraná River has a well-developed multichannel pattern, sand bars are very common in its main and secondary channels. Basically, there are two types of sand bars: central and lateral bars. The first type originates from bedforms (dunes, macro ripples and ripples) that accumulate during large flood events and that emerge when water levels fall, forming bars. Stevaux (1994) described central bars with heights of up to 13 m from the channel bottom and extending more than 1 km in length and with an emerged area of 30,000 m^2, calling these large and complex bed forms '*sand wave*'. The permanence of a central bar in the system is a function of the hydrological regime (i.e. the interval between two large floods) and of the vegetal cover. Normally a central bar remains in the channel for 1 to 10 years before it is eroded (Stevaux and Latrubesse, 2017). However, under some conditions, central bars can evolve over a longer period of time to become an island (Figure 3.4).

Lateral bars with varied morphology are also very common in this section of the Upper Paraná River. Santos et al. (2017) described the processes of lateral bar formation, where they are always associated with islands or the banks of the channel (Figure 3.5, top). Under special flow conditions, a lateral bar can be annexed to an island (or channel border), producing a blind channel (Leli et al., 2018), which is a peculiar morphology, with ecological importance (Figure 3.5, bottom).

The islands play a very important role in river ecology, especially in the distribution and development of riparian vegetation. Palinomorphs and seeds in sediments of islands show that these morphologies, in their sedimentary evolution, shelter different floristic compositions during the sedimentary evolution, which change over time in a close relationship of channel-vegetation connectivity (Stevaux et al., 2013). Although the multichannel pattern presents a large number of small channels with sandy bars and islands, the navigation in this stretch

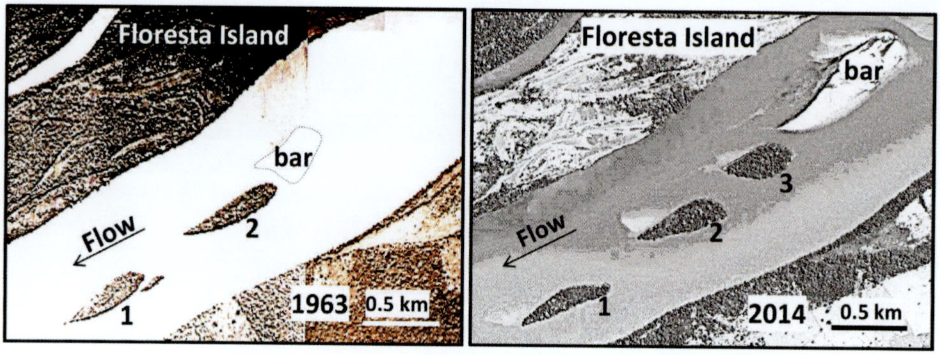

Figure 3.4 (*left*) Aerial photo from 1963 showing a central bar formed upstream of islands 1 and 2, central bar b. (*right*) Satellite image of the same location from 2017 showing the evolution of an ancient central bar to an island 3, and the formation of another central bar and a satellite image upstream. Figure modified from Leli (2015).

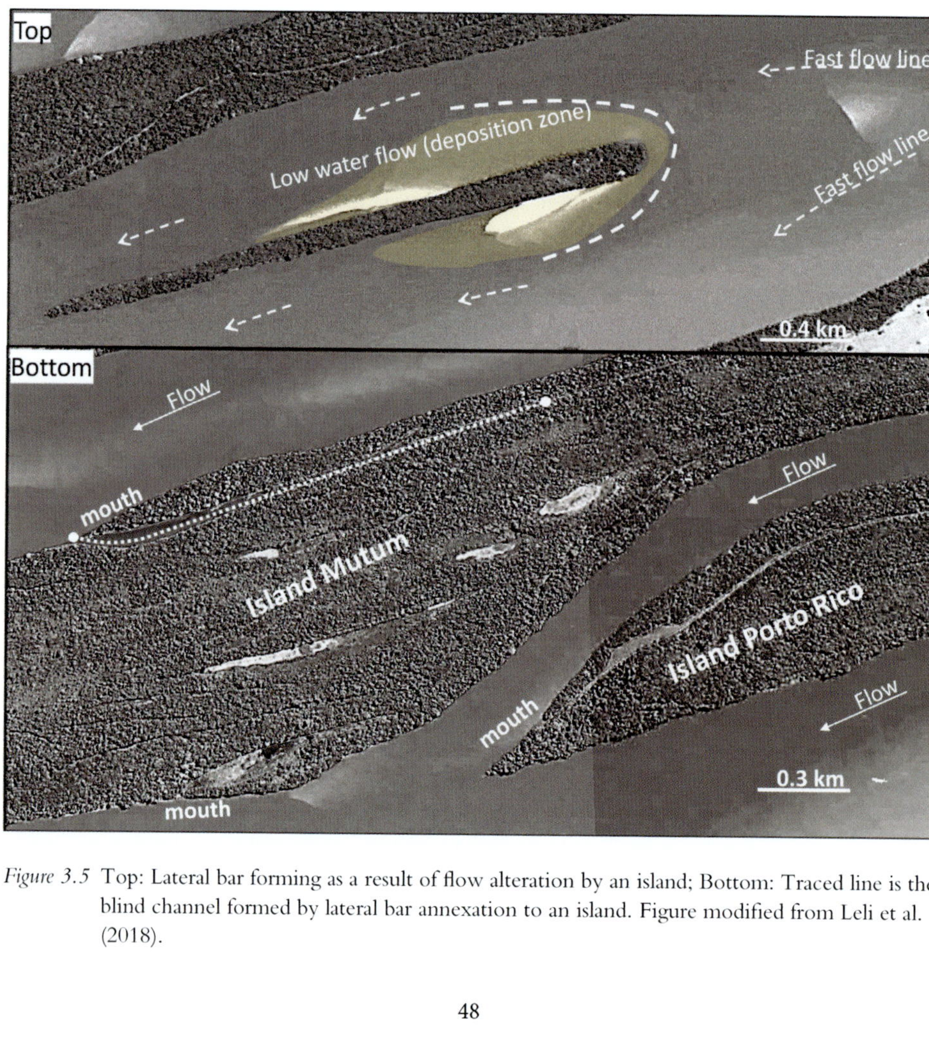

Figure 3.5 Top: Lateral bar forming as a result of flow alteration by an island; Bottom: Traced line is the blind channel formed by lateral bar annexation to an island. Figure modified from Leli et al. (2018).

is made exclusively along the main channel that is located almost entirely on the eastern side of the river. The present hydro-sedimentary dynamics, although altered by the upstream dams, allow navigation along this stretch without the intervention of dredging. In addition, about 1,200 tonnes of sand for civil construction are excavated daily from the river, without producing significant impacts on the river system (Santos et al., 2015).

About 150 large dams with heights >15 m have been constructed in the Upper Paraná River and its main tributaries (Agostinho et al., 2004a, 2004b). Stevaux et al. (2009) described the extreme modifications in suspended sediments, bedload grain size, bed-form morphology, and river regime that occurred over a four-decade period of hydroelectric development. These authors documented a reduction in the mean concentrations of suspended sediments from 35 mg L^{-1} during the pre-dam period, to 24 mg L^{-1} at the beginning of large dam construction (1988–1999), to 14.75 mg L^{-1} (1995–1999) during the construction of the Porto Primavera Dam to 10.85 mg L^{-1} at the completion of the Porto Primavera Dam (with the largest reservoir in the system), to the situation today, with an incredibly low suspended sediment concentration of 0.3 mg L^{-1}. As described in Chapter 8 in this book, these changes have had a great influence on water quality conditions within the dam reservoirs and the downstream transport of nutrients.

In turn, the operation of the Porto Primavera Dam has altered the natural to the regulated discharge profile, with a reduction of the maximum discharge (Qmax) and an increase in the minimum discharge (Qmin). This flattening of the hydrograph has produced a strong impact on floodplain connectivity (Stevaux et al., 2013). Riverbed load was also affected, with a change in the particle diameter of 50% of bed material (i.e. D_{50}) that reflects a change from medium to coarse sand. Changes in the bed load as a result of the altered flow depth modified the size and height of the dunes that form on the river bed (Figure 3.6).

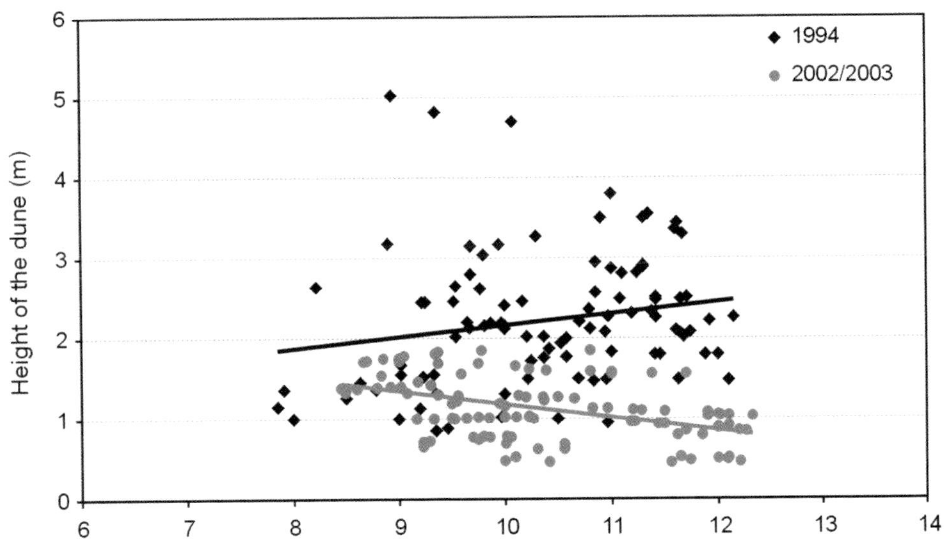

Figure 3.6 Relationship between flow depth and riverbed dune height along the Upper Paraná River thalweg for the pre-operational (1994) and post-operational (2002/2003) periods of the Porto Primavera Dam. Figure modified from Stevaux et al. (2009).

Middle and Lower Paraná River

Downstream of the basaltic plateau at the Brazil-Argentina border, the Paraná River flows into the extensive Argentine plain. In the first stretch, the river generates a large fluvial fan (or megafan), which covers the north-western portion of Corrientes Province in Argentina and the southern part of eastern Paraguay. The megafan is 260 km long and 500 km wide in the distal part and has been stable for most of the Upper Quaternary period. Through a sequence of events over thousands of years, the main channels were eventually replaced by smaller channels and large swamps, leading to the formation of the present megafan. A few extensive deposits produced by spills during dry climatic phases in the Upper Quaternary are intercalated among the abandoned belts. Also, during dry phases, important deflation of sand occurred in the abandoned belts, generating dune fields up to 80-km-long and 5-km-wide. Most of these fluvial deposits in Argentina are named Ituzaingó Formation, and they represent the typical sedimentation pattern of the Paraná River during its evolution. The present river belt crosses the fan in an east-west direction. The channel design there is multi-channelled, with frequent elliptical islands and sand bars.

During the Pliocene, the Iberá Region located in the north-eastern part of Corrientes Province was a wide floodplain of the river, characterized by low-sinuosity channels separated by sandy natural levees. Movements within the Earth's crust owing to the Andean tectonics modified the sub-surface structure, changing the regional slope and drainage direction of the surface waters (Orfeo and Neiff, 2008). The resulting river, now known as the Paraná, gradually abandoned its earlier channel in the region, currently occupied by the Esteros del Iberá and the Corriente River. Thus, the abandoned floodplain remained covered by numerous lakes that nowadays persist due to the influence of rains and possibly infiltration of groundwater coming from the Yacyretá dam, located towards the headwaters of the Iberá. Covering more than 30,000 km^2, the Iberá Region supports a great biodiversity and constitutes one of the largest tropical wetlands in South America (Orfeo and Neiff, 2008). The youngest abandoned belt in the megafan is from the Holocene age and runs along the eastern limit of the system and includes some of the lakes presently located in the Iberá wetland.

Paleocurrent directions measured in the Ituzaingó Formation at Corrientes Province suggest that there was a historical expansion and downstream migration of the Paraná River. A comparison between older and present fluvial deposits reveals that both are homologous, but the first ones seem to be smaller than those belonging to the main channel of the present river, which could be related to the sedimentation of lateral and central channel-bars. In comparison, the present deposits exhibit sedimentological features linked with higher energy conditions compared with the older paleo-deposits (Orfeo and Iriondo, 2010).

The highest discharge of the Paraná River at Corrientes City was 60,215 $m^3 s^{-1}$ measured in the main channel in 1983 (Orfeo and Stevaux, 2002; Neiff et al., 2000). Here, the annual suspended sediment transport within the Paraná River is about 118.7 10^6 tons, and that represents around 75% of the total solids discharge of the entire system (Orfeo and Stevaux, 2002). The Paraná River is the most important waterway in Argentina for the export of fluvial materials and is one of the three most important in South America after the Amazon and Orinoco. There is a close relationship between the flows and transport of sediments in the river and the physical, chemical and ecological characteristics of the basin.

Suspended sediment

The Bermejo River exports a huge volume of fine sediment coming from the Andes Mountains to the Paraguay River that is received by the Paraná River near Corrientes City.

From the confluence between the Paraguay and Paraná rivers, the river shows pronounced spatial and temporal differences in the concentrations of suspended sediment that are related to the discharge of both rivers. Before this confluence with the Paraguay River, the size range of the suspended sediment in the Paraná is homogeneous, represented by clay (81.5%) and silt (18.5%). Downstream of the confluence, the sediment near the eastern margin maintains the aforementioned composition, while on the opposite margin the sediments are dominated by coarser fractions, including very fine sand in suspension. In quantitative terms, the suspended sediment concentration in the Paraná River downstream of the confluence with the Paraguay River shows extremes from 1 to 616 mg L^{-1} on the eastern margin and from 8 to 1,271 mg L^{-1} on the western margin (Table 3.2).

In general, there is a direct relationship between the concentrations of suspended sediments and the discharge of the Middle Paraná River, where the values of the western margin are always higher compared with the eastern margin, owing to the influence of the Paraguay-Bermejo river system (Table 3.2). However, there are short-term variations in the concentrations that deviate from the annual pattern, which can be explained by the influence of the tributary network.

Comparing samples of sediment collected in a cross-section from the western to the eastern margins, there is a higher proportion of coarse particles (>3 μm) along the western margin. Moreover, the proportions of clay and colloids in samples from the eastern margin are two times and three times higher than the samples from the opposite margin, respectively. Despite the relatively low proportion of colloids, this fraction represents a yearly transport of higher than 200,000 tonnes $year^{-1}$, which is significant in terms of the transport of nutrients, metals and organic material (Orfeo and Iriondo, 2002). The proportions of coarser particles (i.e. >3 μm) maintain high fidelity with hydrological changes in the Paraguay River. In comparison, the proportions of finer fractions are not strictly related the hydrometric levels in the rivers.

Due to these differences in suspended sediment concentrations, the mixing interface between the waters can be clearly visualized (Lane et al., 2008; Parsons et al., 2004). The confluence is dominated by a large scour that is ~30 m deep, with its deepest location being within the Río Paraguay due to the local bedrock, but the scour extends into the downstream Paraná River channel. Flow is accelerated as the two flows combine and produces a series of Kelvin-Helmholtz shear layer instabilities that are visible due to the stark contrast in the suspended sediment concentration.

Table 3.2 Suspended sediment concentrations (mg L^{-1}) in the Middle Paraná River at the city of Corrientes, from 2001–2010.

	Eastern Margin			*Western Margin*		
Year	*Minimum*	*Maximum*	*Annual average*	*Minimum*	*Maximum*	*Annual average*
2001	7	311	49	42	828	252
2002	5	584	75	40	908	301
2003	2	298	41	8	778	252
2004	3	202	32	47	768	244
2005	1	41	11	10	526	194
2006	1	158	23	31	1271	312
2007	3	382	82	104	1128	642
2009	2	616	96	32	1154	427
2010	2	180	30	30	1158	267

The controls on the mixing processes at the Paraguay-Paraná confluence, as well as others in the basin, have been the subject of recent investigations. The lack of coherent, channel-scale secondary flows has been attributed to the often slow mixing at the site, with shear-generated turbulence often not being strong enough to mix the two flows (Lane et al., 2008). These differences, including those in the sedimentary size fractions that occur on the two sides of the river are often preserved more than 400 km downstream of the confluence over the annual hydrological cycle, with slight homogenization downstream owing to the progressive mixing process. However, under certain combinations of momentum ratio between the two flows at the confluence, channel-scale secondary flows can form and produce very rapid mixing. A rapid mixing process may be completed in only an 8-km stretch downstream of the junction, depending on the bed morphology at channel junctions. Different mixing rates at the confluence of rivers under the control of basin-scale hydrological cycles are more likely to occur in large river systems than in small rivers (Lane et al., 2008).

The clay minerals suspended in the Middle Paraná River are illite, caolinite, chlorite and smectites. Before the confluence with the Paraguay River, the dominant minerals are chlorite and caolinite (between 80% and 90% for both). The rest corresponds to illite (20%) on the western margin owing to the influence of the Bermejo River, and smectites (10%) on the opposite margin, from the influence of the Upper Paraná (Bonetto and Orfeo, 1984).

An unusual type of sediment transport is recorded in the roots of an aquatic macrophyte, *Eichhornia crassipes*, which is one of the most abundant plants in the water bodies located in the Paraná River floodplain along the western margin. At the high-water period, the inorganic sediment retention by the roots of *E. crassipes* is higher than 1,326 g m^{-2} in floating meadows of this plant that occur in oxbow lakes near the main channel of the Paraná River. The mineralogy of the clay fraction retained in the roots has a consistent composition, with illite being the dominant mineral (near 70%), followed by chlorite (20%) and kaolinite (5-20%), as reported by Poi de Neiff et al. (1994). During the peak of floods, a higher proportion of sand is detected in the roots. Floodplain wetlands covered by *E. crassipes* accumulate and transfer organic and inorganic materials within the flood valley, especially during high-water periods. Thus, any organic or inorganic nutrients captured by the meadows during normal floods represent net increments for the lake (Poi de Neiff et al., 1994).

Taking into account the clay mineral association transported in suspension by the Paraná or retained by the roots of aquatic plants in the surrounding waterbodies, it can be inferred that kaolinite and smectites are probably supplied by the Paraguay and Upper Paraná rivers that drain the tropical Brazilian shields and extensive areas covered by basalts and lateritic soils (Bonetto and Orfeo, 1984; Depetris and Griffin, 1968). Conversely, illite is supplied by the Paraguay-Bermejo River system, which drains the arid environments of the Andean mountains.

Bed sediment and bedforms

The bed sediment of the Paraná River channel in the Middle stretch is predominantly medium (60–80%) and fine (20–40%) sand, with the median grain size of samples ranging between 0.18 mm and 0.99 mm, and an overall mean of 0.38 mm. Most of the bed of the Río Paraná is dominated by dune bed-forms. This bed sediment comes mainly from the Upper stretch of the Paraná (97%). In contrast, most of the suspended sediment in this river (52%) is supplied by the Bermejo River through the Paraguay River (Orfeo and Stevaux, 2002). The deposition of sand travelling in suspension close to the bed contributes to dune migration and may impart considerable control on the lee-side slope angle (Kostaschuk et al., 2009). From Ground Penetration Radar (GPR) surveys over a sandy bar near the confluence between the Paraná and Paraguay rivers, Sambrook

Smith et al. (2009) showed that the deposits are dominated (80% of the facies) by trough and planar cross-strata associated with the migration of dunes. The thickness of these sets decreases from the base to the top of the profiles, and this is associated with the decrease in their formative flow depths. The other major facies present are high-angle strata associated with sediment accretion at bar margins or bar-tails (15% of all facies). This type of facies is much less prevalent in medial areas of the bar, giving rise to a high degree of lateral and downstream variability in the bar deposits.

In the Lower Paraná River, the characteristic sandy bed is maintained but the grain size shows a shift to fine sand. Thus, there is a clear homogeneity in the bed sediments along the Middle and Lower reaches of the Paraná River. In this last stretch, the largest dunes are smaller than in the Middle stretch with maximum and average wave heights of 7 m and 1.5 m, respectively (Marchese et al., 2002)

Channel pattern and morphology

The present multichannel pattern of the Middle Paraná River is related to the abundance of the bedload discharge, which is higher than 25% of the total sediment discharge. Because of this and the regional slope, several kinds of bars and channels are present in the Middle Paraná as a typical feature of this fluvial model. Greater than 32% of the Paraná River channel at Corrientes City is occupied by bars, with surface areas ranging from 0.04 km^2 to 35 km^2. According to the relationship between the subaerial exposition of the bar surface and the water level, these bars may be classified as either islands, if they remain emergent during ordinary floods, or emerging bars if they are flooded every year. The bars also can be generically grouped into 'aggradation bars' when they come from the accumulation of load material, and 'relictual bars' when they are formed from the erosive processes that isolate portions of the alluvial-plain inside the channel (Figure 3.7). In this last case, the relictual bars are essentially part of an old alluvial-plain, abandoned by the river during the lateral migration. In the Paraná River, it is still possible to recognize some portions of the paleo-floodplain of the Paraguay River (Orfeo, 1998). This is a consequence of the upstream migration of this major confluence. Thus, it is possible to identify some ancient Paraná-Paraguay fluvial confluence areas, the southernmost of which is located 40 km to the south-west of the present confluence.

Sambrook Smith et al. (2009) carried out a GPR survey of an aggradation central bar with depth of penetration up to 12 m below the bar surface. Two facies were found to dominate the sedimentary architecture. The principal facies (~83% of total facies) comprises trough and planar cross-strata related to the migration of dunes, with the thickness of the cross-strata decreasing towards the bar surface. The second significant facies (~15%) is high-angle (generally 10°–20°) strata that typically form by accretion at the bar margins or the extreme downstream of the bar. Clay drapes (<2%) and crossbar channels (<1%) were found to comprise only a minor part of the deposits.

In this area, the Middle Paraná River flows across a region composed of several structural blocks elongated in the north-south direction, most of which are tilted slightly to the east. From the Paraná-Paraguay confluence, the Paraná flows 75 km on a subsiding block, widening the floodplain from 13 to 45 km (Iriondo, 1988). For that reason, the floodplain in this region is asymmetrical, developed primarily on the west side of the river due to tectonic tilting of Cretaceous basaltic bedrocks. These formations are related to the emergence of the Andes Mountains during the Cenozoic age. In consequence, the east margin of the Paraná River at the city of Corrientes is about 8 m higher compared with the right margin, and the general flow direction is east-west and north-south, upstream and downstream of Corrientes, respectively.

Figure 3.7 Two types of bars formed in the Middle Paraná River downstream of the confluence with the Paraguay River near the cities of Resistencia and Corrientes (Argentina). A = aggradation bars; R = relictual bars.

In this part of the Middle Paraná River, it is possible to recognize two main units in the floodplain: (a) a proximal floodplain that is directly connected with the main channel, and (b) a distal floodplain (Figure 3.8). In a cross section of the river valley in this location, the distal floodplain is topographically higher and therefore, flooded with less frequency. In general, the main channel of the river covers around 20% of the total fluvial surface in this area of the basin. The rest is occupied by the floodplain (proximal 35%; distal 45%). Upstream and downstream of Corrientes City, respectively, the width of the proximal floodplain ranges from 0.4 to 3 km (mean width of 1.9 km) and from 0.3 to 20 km (mean width of 8.1 km). This elongated and narrow area in contact with the main channel of the Paraná is affected almost annually by the ordinary floods of the river. The distal floodplain is wider than the proximal one (mean value 12.6 km; maximum 21.4 km). During extraordinary floods, the water reaches this region, so that the entire flooded area measures 20.7 km in an average width (ranging from 9.6 km to 26.1 km).

Small tributaries

The headwaters of the small tributaries in this part of the Paraná River Basin are located in large wetlands (Figure 3.7). The channel beds are covered by fine alluvial sediments consisting of silt and cohesive clays (Orfeo, 2006). Some of these tributaries have incremental increases in the sand content and mean size of bed sediments in the flow direction. This sandy sediment

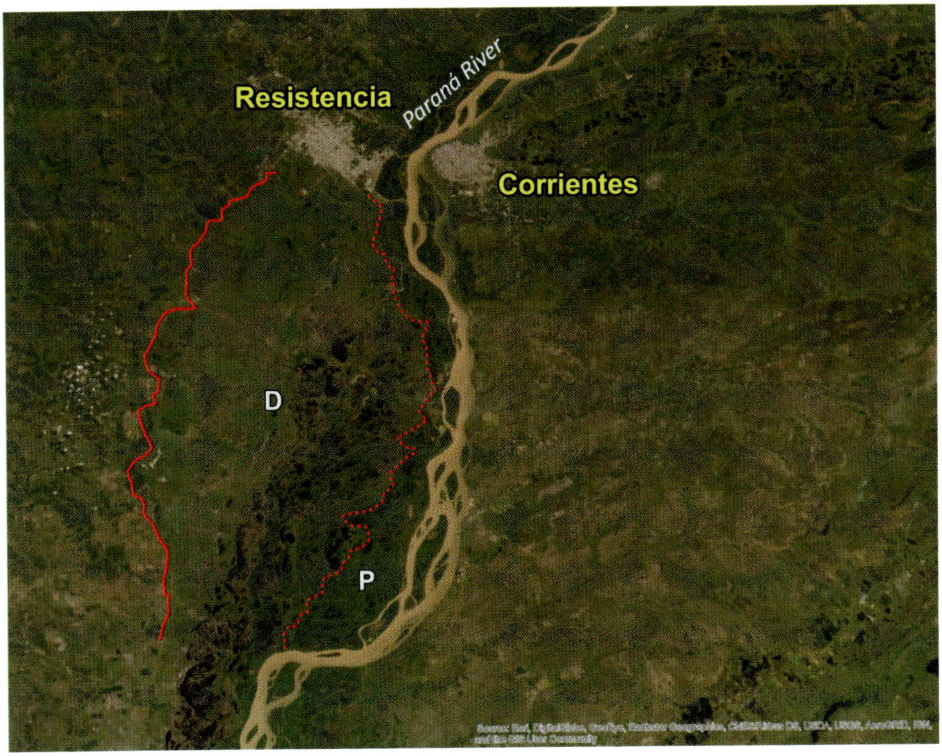

Figure 3.8 Middle Paraná River floodplain downstream of the cities of Resistencia and Corrientes (Argentina). P = proximal floodplain; D = distal floodplain. The dashed red line is the border between the proximal and distal floodplain, and the solid red line is the geographic limit of the fluvial valley.

was deposited previously by the Paraná River. Typically, these types of rivers experience the following hydrological cycles:

- Low water period: the fluvial system is reduced to its minimal condition, facilitating the sedimentation of fine material (<4 μm). Most of the channel and neighbouring areas are rapidly colonized by aquatic vegetation, bringing abundant organic matter that is added to the incoming organic matter from the wet highlands. The high concentration of suspended sediment corresponds to the minimal discharge, and therefore the overall sediment transport is low.
- Beginning of the flood period: The rainfall mobilizes this autochthonous fluvial network, generating the movement of organic and inorganic material deposited on the bed channel and adjacent areas. The suspended sediment concentration increases in concert with the discharge, so that the total solid transport is very high.
- High water period: Most of the bed channels remain free of non-cohesive accumulation, which is exported as wash load to the Paraná River. The discharge can reach a high magnitude, but the suspended sediment concentration is often low due to depletion in supply. At the headwaters, the exposed bed is often characterized by a cohesive clay base. However, in the lower stretches of the rivers, the channel is sometimes covered by clean sand when paleo-deposits of Paraná River are exposed.

Impacts of dams

The construction of megadams on large rivers poses considerable impacts to the environment, such as ecosystem fragmentation, habitat changes, hydrologic alterations of both the quantity and timing of river flow, downstream sediment starvation, downstream changes in water quality, relocation of animal and human populations, species extinctions and the release of greenhouse gases from decaying vegetation. Likewise, the Paraná accumulates more than 10^3 tonnes yr^{-1} of total plastic loads, which show a strong correlation with population density, urbanization and the degree of wastewater treatment (Best, 2019).

The hydro-sedimentology of the Paraná River in the Middle and Lower stretches has been modified by natural and human impacts that have altered the transport of solids in the river and almost all its tributaries. When considering the present suspended sediment concentration of the Paraná River in the middle stretch with those of the '80s and '90s, a decrease in solids discharge has been observed, in spite of the overall increase in discharge since the 1970s (Amsler and Drago, 1999). This observation is clearly related to the construction of dams in the main channel and some of the tributaries of the Paraná River. The last structure is the Yacyretá Dam, which was constructed as a bi-national project between Argentina and Paraguay. The sediment retention capacity of those dams is particularly important, considering the increase in anthropogenic influences in the region, with greater urbanization and industrial development, intensive agriculture, deforestation, etc. (Amsler, 2006).

The Paraná River channel is a navigation route for trade, as well as a source of river sediments for construction materials. The need to ensure passable channels for navigation through dredging may impact hydrologically sensitive regions, and especially wetlands such as the Pantanal located in the Paraguay River basin, that could be adversely affected by dredging or flow diversions. Moreover, sediment mining may contribute significantly to a decreasing downstream sediment flux, which may exacerbate the effects of subsidence in river deltas, as well as causing scour around in-channel engineering structures and potentially triggering channel incision and bank failures (Best, 2019).

The Delta of the Paraná River

The Delta region of the Lower Paraná is located from 32°S to 34°S, beginning downstream of the city of Paraná in Entre Ríos Province, Argentina, where the river divides into several branches (Figure 3.9). The area has an altitude above sea level of less than 10 m, with a gentle slope to the southeast. Once in the deltaic area, the branches and minor tributaries adopt the typical fluvial design. The subaerial surfaces are wide flooded lands, limited by meandering and anastomosed streams. Scroll meanders, point bars and oxbow lakes linked with abandoned meanders, are also frequent (Pereyra, 2003).

This littoral complex developed during the Holocene (Iriondo, 2005), initiating with a fluvial phase, probably in the early Holocene. During the middle Holocene, this region was affected by a marine ingression that reached up to the area currently occupied by Rosario City in Santa Fe province, resulting in the accumulation of shells and silty materials near the littoral margin at the southern margin of Entre Ríos province. Those coastal ribbons controlled the drainage design of the tributaries, where old tidal channels can also be recognized. Over the tidal plains, some field dunes that are partially active at present were developed (Pereyra, 2003).

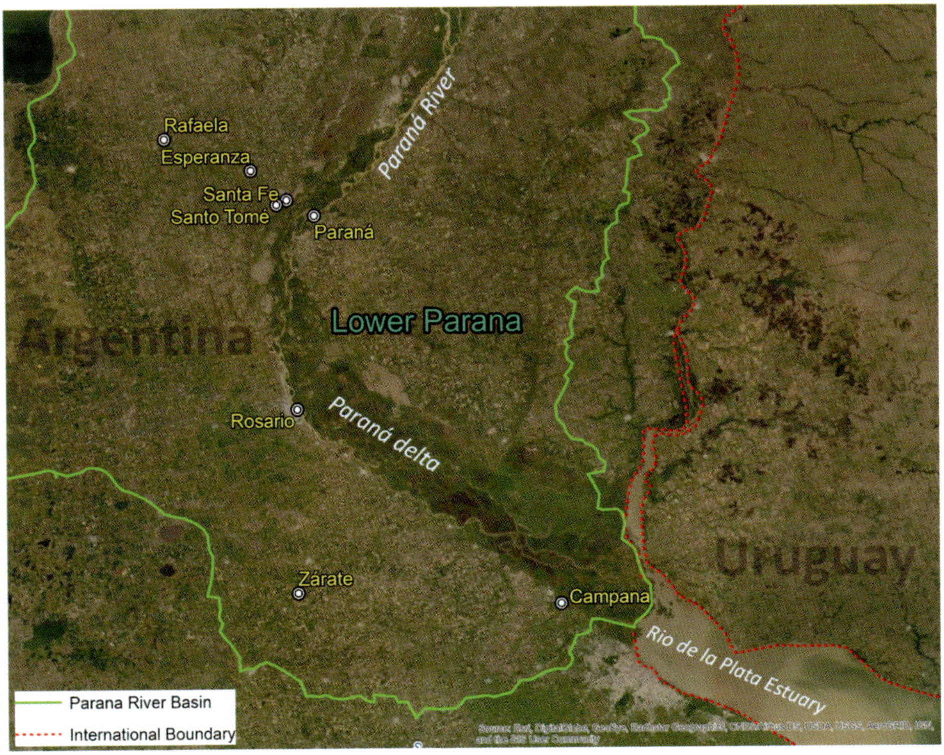

Figure 3.9 The Delta region of the Lower Paraná River showing the major urban centers and the limits of the river basin. The dotted red line indicates the border between Argentina and Uruguay.

The Delta has a typical triangular shape and was built by the frontal advance of the Paraná into the Río de la Plata, where wide channels with well-developed natural levees from 70 m to 120 m in width composed of fine sand, are distinctive features. The central part of these islands is occupied by shallow lakes or marshes where the deposition of silt and clay transported in suspension beyond the natural levees is dominant. The Paraná River discharges most of its sedimentary load at the bottom of the frontal area of the Delta, with a delta front progradation rate between 42.3 m yr^{-1} and 45.1 m yr^{-1}, as measured since 1933 (Marcomini et al., 2018). Here, the 1 m tidal amplitude plays a significant role in the distribution and deposition of the sediments. At present, this sedimentological unit is 90 km in length with a surface area of 3,500 km^2, but models indicate that the surface area of the Delta could increase to about 3,914 m^2 by 2050 (Marcomini et al., 2018).

The recent rapid growth in the human population within the Delta region has resulted in an increase in the price of land, as well as increasing the rate of exploitation of natural resources. The Paraná Delta front has moved episodically towards the estuary, with adjoining distributary mouth bars increasing, resulting in expansion of the subaerial surface of the lower delta plain. This rapid progradation has changed several sub-environments and ecosystems; not only along the Delta margin, but also along the estuary coast. Therefore, it is very important in the future to predict, plan and legislate the pattern of occupation of these new territories to avoid future social, economic, political and cultural conflicts (Marcomini et al., 2018).

Conclusions

The regulated areas of the Paraná River are primarily located in Brazil and in the northern part of Argentina along the border with Paraguay. Losses of soils as a result of erosion from agriculture and forestry practices are major factors affecting transport of sediments within the system. This process may not only alter the distribution of the sediments in the channels, but also could trigger the erosion of the banks and other important morphological changes like lateral migration. Alterations to the hydrological and sedimentological conditions within the river can influence negatively the survival of animal or plant species living in the ecosystems within the Paraná River Basin.

There are drastic differences in the concentrations of suspended sediments between the Upper stretch and the Middle-Lower stretches of the basin. In the Brazilian highlands, the concentrations of suspended solids are very low owing to the influence of the numerous dams built in the main channel and tributaries that retain most of the wash load and also part of the bed load. At the confluence with the Paraguay River, the Paraná receives huge quantities of sediment in suspension coming from the Andes Mountains that are discharged by the Bermejo River into the lower Paraguay. This suspended sediment discharge is not deposited in the channel of the Paraná until it arrives at the deltaic area in the Lower Paraná.

The Middle Reach of the Paraná River is characterized by a multichannel pattern due to the high frequency of discharge of its bedload sediment, which is composed mainly of sand. Therefore, the main channel is characterized by a large number of sand bars and vegetated islands that distribute the flow into several branches. The floodplain in the Middle Paraná has extensive wetlands, shallow lakes and small tributaries. Its location is asymmetrical due to tectonic control that originated with the movement of basaltic blocks on the western margin of the river. For this reason, the floodplain is not developed along the eastern bank, owing to the different relative altitudes.

The geomorphological and sedimentological patterns within the various stretches of the Paraná River Basin have a profound effect on the distribution of urban centers within the basin, the location of agriculture and livestock operations, fishing, navigation and hydroelectric power generation. The natural habitats that support the extensive biodiversity within the basin are dependent on the sedimentological characteristics of the various reaches of the basin. Finally, the quality of the water for potable use and for irrigation is dependent on the changing sediment loads over the course of the river.

Acknowledgements

Many thanks to Erin Hayward for the preparing maps and figures using ArcMap. We are also greatly thankful to the co-editor Professor Chris Metcalfe of Trent University (Ontario, Canada) and the anonymous reviewers for their valuable comments and assistance for the improvement of this paper.

References

Agostinho, A.A., Gomes, L.C., Veríssimo, S., Okada, E.K. 2004a. Flood regime, dam regulation and fish in the Upper Paraná River: Effects on assemblage attributes, reproduction and recruitment. *Reviews in Fish Biology and Fisheries* 14:11–19.

Agostinho, A.A., Thomaz, S.M., Gomes, L.C. 2004b. Threats for biodiversity in the floodplain of the Upper Paraná River: Effects of hydrological regulation by dams. *Ecobiology and Hydrobiology* 4:255–268.

Amsler, M., 2006. Evolución de la carga de lavado en Alto Paraná (1968-2004). Incidencia sobre las sedimentaciones en la planicie aluvial del Paraná Medio. *Proc. III Congreso Iberoamericano sobre Control de la Erosión y los Sedimentos*, Buenos Aires, Argentina.

Amsler, M., Drago, E., 1999. A review of the suspended sediment budget at the confluence of the Paraná and Paraguay Rivers. *Proc. Hydrological and Geochemical Processes in Large Scale River Basins*, Manaus, Brazil.

Best, J. 2019. Anthropogenic stresses on the world's big rivers. *Nature GeoScience* 12:7–21.

Best, J., Ashworth, P., Bridge, J., Lane, S., Lunt, I., Orfeo, O., Parsons, D., Sambrook Smith, G., Simpson, C. 2006. The four-dimensional sedimentology of sandy braided rivers: Linking process to product, assessing the impact of flood magnitude and questions of scale invariance. *Proc. IV Congreso Latinoamericano de Sedimentología – XI Reunión Argentina de Sedimentología.*

Bonetto, A., Orfeo, O. 1984. Caracteres sedimentológicos de la carga en suspensión del río Paraná entre Corrientes y Esquina (Argentina). *Revista de la Asociación Argentina de Mineralogía, Petrología y Sedimentología* 15:51–61.

Depetris, P., Griffin, J. 1968. Suspended load in the Río de la Plata drainage basin. *Sedimentology* 11:53–60.

Fortes, E., Stevaux, J.C., Volkmer, S. 2005. Neotectonics and channel evolution of the Lower Ivinhema River: A right-bank tributary of the upper Paraná River, Brazil. *Geomorphology* 70:325–338.

Gon, P.P. 2012. *Morfologia e hidrodinâmica em rios multicanais ("anabranching"): exemplo do alto rio Paraná.* Dissertação de mestrado, Universidade Estadual Paulista, Rio Claro, São Paulo, Brazil.

Iriondo, M. 1988. A comparison between the Amazon and the Paraná fluvial systems. *Mitteilungen des Geologisch-Paläontologisches Instituts der Universität Hamburg, SCOPE/UNEP Sonderband, Hamburg, H.* 66: S77–S92.

Iriondo, M. 2005. El complejo litoral en la desembocadura del río Paraná. *Proc. XVI Congreso Geológico Argentino, La Plata*, Buenos Aires, Argentina.

Kostaschuk, R., Shugar, D., Best, J., Parsons, D., Lane, S., Hardy, R., Orfeo, O. 2009. Suspended sediment transport and deposition over a dune: Río Paraná, Argentina. *Earth Surface Processes and Landforms* 34:1605–1611.

Lane, S., Parsons, D., Best, J., Orfeo, O., Kostaschuck, R., Hardy, R. 2008. Causes of rapid mixing at a junction of two large rivers: Río Paraná and Río Paraguay, Argentina. *Journal of Geophysical Research – Earth Surface* 113:1–16.

Leli, I.T. 2015. *Gênese, evolução e geomorfologia das ilhas e planície de inundação do Alto Rio Paraná, Brasil.* (Ph.D. Thesis). Department of Geoscience. State University of São Paulo, Rio Claro, São Paulo, Brazil.

Leli, I.T., Stevaux, J.C., Assine, M.L. 2018. Genesis and sedimentary record of blind channel and islands of the anabranching river: An evolution model. *Geomorphology* 302:35–45.

Marchese, M., Ezcurra de Drago, I., Drago, E. 2002. Benthic macroinvertebrates and physical habitats relationships in the Paraná River flood-plain system. In: McClain, M. (ed.), *The Ecohydrology of South American Rivers and Wetlands*, IAHS, Special Publication of the International Association of Hydrological Science, Potsdam, Germany, 6, pp. 111–130.

Marcomini, S., Tripaldi, A., Leal, P., López, R., Alonso, M., Ciccioli, P., Quesada, A., Bunicontro, P. 2018. Morphodynamics and sedimentation of a sector of the Paraná delta front between 1933 and 2016, Province of Buenos Aires, Argentina. *Revista de la Asociación Geológica Argentina* 75:1–16.

Martins, D.P., Stevaux, J.C. 2005. Formas de leito e transporte de carga de fundo do alto eio Paraná. *Revista Brasileira de Geomorfologia* 6:43–50.

Neiff, J., Mendiondo, E., Depetris, C. 2000. ENSO floods on river ecosystems: Catastrophes or myths?. In: Toenmsnann, F., Koch, M. (eds.), *River Flood Defence, Kassel Reports on Hydraulic Engineering 9*, Herkules Verlag, Berlin, pp. 141–152.

Orfeo, O. 1998. Reconstrucción morfológica y paleohidrológica de un tramo abandonado del río Paraguay inferior (Argentina). *Revista de la Asociación Geológica Argentina* 53:348–353.

Orfeo, O. 2006. Dynamics of sediment transport in two subtropical plain rivers of South America. *Zeitschrift für Geomorphologie* 145(Suppl.):229–241.

Orfeo, O., Iriondo, M. 2002. *Variación quincenal de la concentración y tamaño de sólidos suspendidos del río Paraná (Corrientes, Argentina)*, Reunión de Comunicaciones Científicas y Tecnológicas de la Universidad Nacional del Nordeste, Argentina.

Orfeo, O., Iriondo, M. 2010. The megafan of the Paraná River from the Pleistocene to the present. *Proc. 18th International Sedimentological Congress*, Mendoza, Argentina.

Orfeo, O., Neiff, J. 2008. Esteros del Iberá: Un enorme laboratorio a cielo abierto. In: Sitios de interés geológico de la República Argentina, Servicio Geológico Minero Argentino, Buenos Aires, Argentina. *Anales* 46:415–425.

Orfeo, O., Stevaux, J. 2002. Hydraulic and morphologic characteristics of middle and upper reaches of the Paraná River (Argentina and Brazil). *Geomorphology* 44:309–322.

Parsons, D., Best, J., Lane, S., Orfeo, O., Kostaschuk, R., Hardy, R., Franklin, M. 2004. The morphology and sediment dynamics of a large river confluence: The Rio Parana and Rio Paraguay, NE Argentina. *Proc. Second International Conference on Fluvial Hydraulics.*

Pereyra, F. 2003. Ecoregiones de la Argentina. *Servicio Geológico Minero Argentino, Anales* 37: 182.

Poi de Neiff, A., Neiff, J., Orfeo, O., Carignan, R. 1994. Quantitative importance of particulate matter retention by the roots of *Eichhornia crassipes* in the Paraná floodplain. *Aquatic Botany* 47:213–223.

Sambrook Smith, G., Ashworth, P., Best, J., Lunt, I., Orfeo, O., Parsons, D. 2009. The sedimentology and alluvial architecture of a large braid bar, Río Paraná, Argentina. *Journal of Sedimentary Research* 79:629–642.

Santos, D.N., Saad, A.R., Stevaux, J.C. 2015. Mineração de Areia em Canais Fluviais de Grandes Rios Tropicais: Análise na Extração de Areia No Alto Curso do Rio Paraná entre os Municípios de Marilena e Guaíra (PR). *Espacios (Caracas)* 36:17–26.

Santos, V.V., Stevaux, J.C., Assine, M.L. 2017. Fluvial processes in attachment bars in the Upper Paraná River, Brazil. *Revista Brasileira de Geomorfologia* 18:483–499.

Souza Filho, E.E., Stevaux, J.C. 2004. Geology and geomorphology of the Baía Curitiba-Ivinheima River complex. In: Thomaz, S.M., Agostinho, A.A., Hahn, N.S. (eds.), *The Upper Paraná River and its Floodplain: Physical Aspects, Ecology and Conservation*, Backhuys Publishers, Leiden, the Netherlands, pp. 1–29.

Souza, M.D., Jr., Santos, M.L., Salamuni, E., Stevaux, J.C., Morales, N. 2013. Analise morfométrica da bacia do rio Ivaí, PR., curso inferior. *Revista Brasileira de Geomorfologia* 14:208–218.

Stevaux, J.C. 1994. The Upper Paraná River (Brazil): Geomorphology, sedimentology and paleoclimatology. *Quaternary International* 21:143–161.

Stevaux, J.C., Corradini, F.A., Aquino, S. 2013. Connectivity processes and riparian vegetation of the upper Paraná River, Brazil. *Journal of South American Earth Sciences* 25:1–9.

Stevaux, J.C., Latrubesse, E.M. 2017. *Geomorfologia Fluvial*, Oficina de Textos, São Paulo, Brazil, p. 320.

Stevaux, J.C., Martins, P.D., Meurer, M. 2009. Changes in large tropical rivers: The Paraná River downstream from the Porto Primavera Dam, Brazil. *Geomorphology* 113:230–238.

4
Water infrastructure

José G. Tundisi and Takako Matsumura-Tundisi

Key messages

- There have been 46 large hydroelectric dams plus more than 100 smaller power plants producing less than 2 megawatts constructed in the upper part of the Paraná River Basin.

- Benefits of the construction of these hydroelectric dams include economic development, high employment, flood control, water security and the availability of water for irrigation in agriculture.
- The reservoirs behind the hydroelectric dams are also used for navigation, public water supply, fisheries, aquaculture, recreation and tourism.
- The transportation system in the Upper Basin (Hidrovia Tietê-Paraná or Tietê-Paraná Waterway) extends over 2,400 km and it is used to transport goods from six Brazilian States. In 2018, 9.7 million tonnes of products were transported.
- Urbanization, deforestation, agriculture and industrial development, applications of pesticides and fertilizers and discharges of untreated wastewater have caused eutrophication, sedimentation and degradation of water quality in the reservoirs.
- Winds with velocities of 10–20 m s^{-1} generated by cold fronts originating from the southern part of South Americ a drive mixing within reservoirs that affects the trophic status of the reservoirs, as well as biogeochemical cycles.
- Eutrophication in reservoirs results in the growth of cyanobacteria (i.e. blue green algae) that produce toxins and degrade water quality so that it cannot be used as a water supply for public use, and is not suitable for fisheries, tourism and recreation.
- Reservoirs are sources of greenhouse gas emissions, and reservoirs with lower hydraulic retention times (i.e. < 20 days) have lower emissions than reservoirs with higher retention hydraulic times (i.e. >100 days).
- However, for reservoirs, the maximum value of CO_2 emissions per kw h^{-1} is 91 g, while for coal and gas thermoelectric plants the value is 930 g of CO_2.
- Ecohydrological approaches that integrate a knowledge of the ecology of reservoirs with Best Management Practices are needed to ensure the sustainability of the basin and the recovery of ecosystem services.
- In the basin, there are examples of regional approaches to managing reservoirs and their watersheds that can be applied to these large and complex systems.

Introduction

The construction of water infrastructure within the Paraná River Basin is mainly related to the development of a massive reservoir system, with the primary objective to produce hydroelectricity. Because of the steep topography in the Upper Basin, all of the hydroelectric plants and reservoirs are confined to the Paraná River upstream of the city of Corrientes in Argentina and to the main tributaries of the Upper Basin, the Rio Grande, Tietê, Paranapanema, Paranaiba and Iguaçu rivers (Tundisi, 1994; Agostinho et al., 2007a, 2007b; Stevaux et al., 2009). Additional infrastructure developments in the Upper Paraná basin include an extensive water transportation system, the Hidrovia Tietê-Paraná (i.e. Tietê-Paraná waterway).

At the beginning of the 20th century, there was an increase in the demand for energy in South America as a result of industrialization and urbanization, especially in the southeast region of Brazil. Therefore, investment in resources and technology for hydroelectricity production and transmission lines promoted rapid progress in the production of hydropower. The first public hydroelectric plant, which was inaugurated in 1889 was the Marmelos plant in the Paraibuna River in Minas Gerais State, Brazil, with a relatively small 250 kilowatts of electricity production. The larger Edgar de Souza hydroelectric power plant in the Tietê River was inaugurated in 1901, with a reservoir of 30×10^6 m^3 and 16 MW h^{-1} of energy production. An intensive period of construction of hydroelectric facilities began at the end of the 1940s. From

1950 to 2000, many larger hydroelectric plants began operations, with energy production at each plant ranging from 400 to 12,600 MW.

Main reservoirs of the Paraná River Basin

Beginning in the late 1940s, there was a cascade of construction of larger hydroelectric dams that produced reservoirs that drastically reduced the lotic (flowing river) environment. Table 4.1

Table 4.1 The 46 major hydroelectric power plants that have been developed in the Upper Paraná River and its main tributaries, the Paranapanema, Tietê, Paranaiba and Iguaçu, including data on their location, year of construction, power generation potential, flooded area of the reservoirs and the state/country in which they are located.

Sub-Basin	*Geographical Coordinates*		*Construction Year*	*Power Generation (MW^{-1})*	*State/Country**	*Flooded Area (km²)*
	Latitude	*Longitude*				
PARANÁ RIVER						
Jupiá	26°23′56" S	52°43′38" W	1974	1,551	MS/SP	1195
Porto Primavera	22°28′31" S	52°57′30" W	2003	1,814	SP	2250
Ilha Solteira	20°23′00" S	51°21′00" W	1973	3,444	SP	1195
Itaipu Binacional	25°24′46" S	54°35′39" W	1984	14,000	Brazil/Paraguay	1350
Yacyretá	27°24′20" S	56°15′15" W	1984	3,100	Argentina/Paraguay	815
TIETÊ RIVER						
Barra Bonita	22°31′11" S	48°32′06" W	1967	141	SP	340
Bariri	22°09′14" S	48°45′13" W	1965	143	SP	63
Ibitinga	21°45′00" S	48°59′30" W	1969	132	SP	xx
Promissão	21°17′52" S	49°47′01" W	1977	264	SP	530
Nova Avanhandava	21°07′06" S	50°12′03" W	1982	347	SP	210
Três Irmãos	20°40′10" S	51°18′00" W	1993	808	SP	785
Edgar de Souza	23°27′16" S	46°54′36" W	1901	16	SP	–
Henry Borden (Billings)	23°52′32" S	46°27′05" W	1926	700	SP	–
PARANAPANEMA RIVER						
Jurumirim	23°12′36" S	49°13′48" W	1962	98	SP	449
Canoas I	22°56′26" S	50°31′02" W	1999	83	SP/PR	23
Canoas II	22°56′19" S	50°15′04" W	1999	83	SP/PR	xx
Capivara	22°39′36" S	51°21′29" W	1978	619	SP/PR	576
Chavantes	23°07′42" S	49°43′59" W	1971	414	SP/PR	400
Ourinhos	23°07′42" S	49°43′59" W	2005	44	SP/PR	118
Piraju	23°09′14" S	49°22′46" W	2003	71	SP	2
Salto Grande	22°54′15" S	49°59′54" W	1958	74	SP/PR	783
Taquaruçu	22°32′30" S	51°59′59" W	1992	554	SP	80
Rosana	22°36′07" S	52°52′20" W	1971	372	SP	220
Paranapanema	23°11′14" S	49°23′03" W	1999	31	SP	2

(Continued)

Table 4.1 (Continued)

Sub-Basin	*Geographical Coordinates*		*Construction Year*	*Power Generation (MW^{-1})*	*State/Country**	*Flooded Area (km^2)*
	Latitude	*Longitude*				
RIO GRANDE RIVER						
Agua Vermelha	19°51'30" S	50°20'35" W	1979	1,396	MG	647
Camargos	21°19'28" S	44°36'58" W	1979	46	MG	73
Furnas	20°40'11" S	46°19'05" W	1963	1,216	MG	1440
Igarapava	16°59'30" S	47°45'29" W	1998	210	SP/MG	37
Itutinga	21°17'35" S	44°37'30" W	1955	52	MG	2
Jaguara	20°01'26" S	47°26'01" W	1971	424	SP	35
Luis carlosBarreto de Carvalho	20°09'11" S	47°16'46" W	1969	1,050	SP	xx
Marimbondo	20°18'11" S	49°11'52" W	1975	1,440	SP/MG	438
Peixoto	20°17'11" S	47°03'48" W	1956	476	MG	250
Porto Colomia	20°07'38" S	48°33'54" W	1974	320	SP/MG	143
Funil	21°08'36" S	45°02'11" W	1969	216	RJ	40
Volta Grande	20°01'54" S	48°13'10" W	1974	380	SP/MG	2850
PARANAIBA RIVER						
Cachoeira Dourada	18°30'10" S	49°29'27" W	1982	658	MG/GO	100
Emborcação	18°27'05" S	47°59'13" W	1982	1,192	MG/GO	473
Itumbiara	18°25'09" S	49°12'55" W	1980	2,082	GO	778
São Simão	19°00'55" S	50°29'48" W	1978	1,710	MG/GO	722
IGUAÇU RIVER						
Segredo	25°47'35" S	52°06'47" W	1992	1,260	PR	81
Salto Caxias	25°32'36" S	53°29'48" W	1999	1,240	PR	–
Foz de Areia	26°00'35" S	51°39'44" W	1970	2,500	PR	167
Salto Santiago	25°37'04" S	52°36'48" W	1980	1,420	PR	208
Salto Ozório	25°32'06" S	53°00'33" W	1975	1,078	PR	51
Baixo Iguaçu	25°30'00" S	53°40'00" W	2018	350	PR	–

*States of Brazil: SP (São Paulo); MS (Mato Grosso do Sul); PR (Paraná): MG (Minas Gerais); RJ (Rio de Janeiro); GO (Goiás).

lists the 46 major hydropower plants constructed in the Paraná River and its tributaries. The reservoirs also have extensive wetlands, some with areas >100 km^2. These are important ecotones in the watershed due to their functions as ecological buffers and high biodiversity. As illustrated in Figure 4.1, the 46 larger dams and their reservoirs that were developed over time in the Upper Paraná basin incrementally increased the area of land flooded within the basin and now the reservoirs inundate a total area of 20,979 km^2, or approximately 1% of the basin area.

The development of these reservoirs is overwhelmingly for the generation of hydroelectricity. For instance, the hydroelectric plants supply around 70% of the electricity consumed in Brazil (Agostinho et al., 2008). Due to exhaustion of the possible locations for construction of

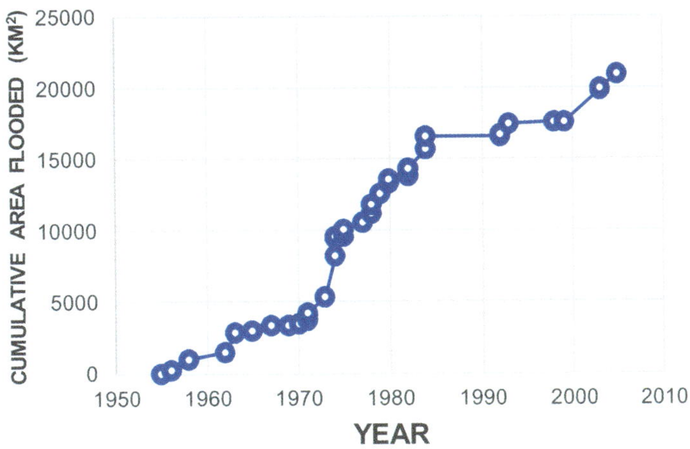

Figure 4.1 Incremental increase in the area flooded in the Paraná River Basin to a total of approximately 21,000 km^2 as a result of the construction of the 46 hydropower plants listed in Table 4.1.

large dams in the Upper Paraná basin, over the past 20 years only two dams were constructed; the Porto Primavera on the Paraná River, which was completed in 2003, and the Yaciretá, also on the Paraná River, which was completed in 2011. The Yaciretá dam, shared by Argentina and Paraguay, was the last dam constructed in the basin, with a capacity of 3.2×10^3 MW of hydroelectric production. Therefore, the construction of large reservoirs in the Upper Basin of the Paraná is effectively over.

The economic development of the Paraná River Basin

The production of hydroelectricity stimulated the economic development of the Paraná River Basin, especially in six Brazilian States. For example, hydroelectric plants located on the Tietê River have a combined capacity for energy production of 2,658 MW. The Itaipu hydropower plant, which is the second largest in the world, produces 14,000 MW of energy and supplies 12–17% of the demand for energy in Brazil. Figure 4.2 illustrates the geographical locations of the 46 large hydroelectric dams and reservoirs in the Paraná River and its main tributaries listed in Table 4.1. The construction of multiple reservoirs has a higher cumulative impact relative to the impact of a single reservoir.

The economic development stimulated by these hydroelectric projects is illustrated by the case of the Hidrovia Tietê-Paraná waterway project, which resulted in construction of 12 terminals for multimodal transportation, 23 industrial centers, 17 touristic centers and 12 distribution centers. Reservoirs of the Tietê and Upper Paraná River have a series of locks that allow navigation by vessels along a stretch of 2,400 km, transporting products from six Brazilian States. In 2018, 9.7 million tonnes of products were transported along this system, including soya beans, cattle, corn, sand and sugar cane. On a monthly basis, a total of >200 ships transporting these products navigate through this waterway. Large boats for tourism also navigate through this complex.

The construction of this water infrastructure has changed considerably the economy of the Upper Basin. The demographic pattern has changed from a primarily rural population in

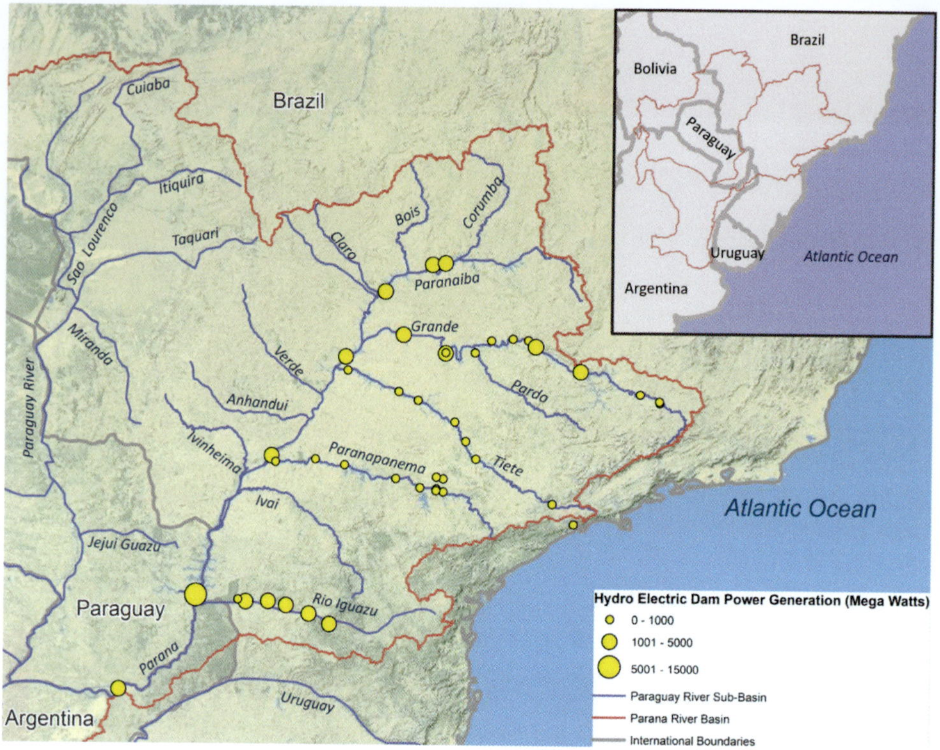

Figure 4.2 Location of the 46 large hydroelectric dams and reservoirs within the Paraná River Basin. The size of the circle represents the hydroelectric energy production of the plant. Hamid Mehmood compiled the geospatial data in ArcMap format and Erin Hayward prepared the map using those data.

1960 (i.e. 77%) to a primarily urban population in 2010 (i.e. 85%), which is a consequence of the new industries, commerce and services in the urban centers. The energy consumption increased from 19,000 MW y^{-1} in 1970 to 443,000 MW y^{-1} in 2010 (Henry, 2014). The construction of reservoirs also had a large impact on agricultural development. Production changed to cultivation of crops with greater market value. New irrigation technologies made possible by access to cheap energy and a permanent water supply increased production. Large-scale cultivation of sugar cane for sugar and bioethanol production and soybeans has increased markedly over the past 30 years. The gross domestic product among the 286 municipalities of the Upper basin increased by 80% over the period from 2000 to 2012 (Valencio and Gonçalves, 2014). The water infrastructure investment was followed by large-scale construction of a network of roads, railroads and gas pipelines in order to improve access to transportation and energy. These conditions stimulated tourism in several reservoirs and their watersheds. The Hidrovia Tietê/Paraná is an important tourist site in São Paulo State that attracts thousands of visitors each year and promotes new jobs, employment and new business (Grananda and Romero, 2014). At least 15 towns are recognized by the Government of São Paulo State in Brazil as special sites for tourism. Another important economic activity developed in the last 20 years in the reservoirs is aquaculture with both native and introduced species.

Ecological dynamics of the reservoirs

Controlling factors

Several studies on the reservoirs of the Paraná River Basin have described the dynamics, the ecology and the limnology of these artificial ecosystems (Tundisi and Matsumura-Tundisi, 2012). The filling stage of the reservoir is a period of intense and rapid transformation, with an abrupt transition from a lotic (i.e. moving freshwater) to a lentic (still freshwater) environment. A trophic upsurge, which results in an increase in the productivity and biomass of some species, is common at this stage, as demonstrated for the reservoirs of the basin where limnological monitoring was performed during the filling phase (Agostinho et al., 1999). The colonization of the reservoirs depends upon the rate of adaptation of the biological communities from a lotic to a lentic environment.

After stabilization of the reservoir dynamics, the regulation and control of the functions of all these artificial ecosystems in the basin is dependent on the factors described below:

- Climatological/hydrological control that is dominant during the period of high rainfall in the Austral summer (i.e. November to March). The high flushing rates and low hydrological retention times control nutrient cycles, seasonal succession of biological communities and primary productivity (Tundisi et al., 2008a).
- Circulation through wind action in the relatively shallow reservoirs of the Upper Paraná basin (i.e. 25 m average depth). Vertical circulation is common to all reservoirs, and therefore they are primarily polymictic ecosystems with only short (i.e. hours or days) periods of thermal stratification (Straskraba and Tundisi, 1999).
- Dynamic fluctuations between riverine, lacustrine (i.e. lakes) and transition zones, resulting in large changes in the horizontal patterns of ecological zoning (Agostinho et al., 2008). Reservoirs are highly unstable systems as a result of these physical, chemical and biological fluctuations.
- Biological forces that impact reservoirs include fish predation (piscivorous and zooplanktivorous), blooms of cyanobacteria as a result of eutrophic conditions during short periods of stratification, changing horizontal distributions of food webs, including fish communities, phytoplankton, zooplankton, invertebrates and phyto-benthos. Populations of migratory fishes decline due to the loss of lentic ecosystems, spatial fragmentation and loss of critical habitats (Agostinho et al., 2008; Henry, 2014).
- Physical, chemical and biological forces in the reservoirs that include width, water flow, turbidity, nutrients, sources of organic matter, trophic status, and the diversity of plankton, non-migratory fish, migratory fish and icthyofauna. All of these processes are distributed horizontally or vertically among the three spatial components of lacustrine, transition and riverine systems (Agostinho et al., 2007b, 2008)
- Operational parameters such as water-withdrawal and water storage that have an effect on water volume, storage schedule and surface elevation. Limnological consequences of these operational procedures include variations in flushing rates and retention times, sedimentation rates and nutrient retention, storage or dissipation of heat, inflow mixing and transport of organic material. Since the reservoirs of the Paraná River Basin are in a cascade series, operational procedures affect all of these ecosystems almost simultaneously (Kennedy, 1999; Straskraba and Tundisi, 1999).

Even considering the variables described above, the ecological and limnological response of the reservoirs to climatological and hydrological forces such as cold fronts, and the discharge resulting from rainfall are remarkably similar. As described below, cold fronts affect the reservoirs by increasing vertical circulation, thus causing large changes in the distribution of nutrients such as phosphorus and nitrogen, as well as suspended particulates and planktonic organisms.

Impacts of cold fronts

The horizontal and vertical structure of the reservoirs depends upon the wind velocity and direction, the intrusion of waters from the tributaries and the retention time. There is strong evidence that cold fronts originating from the southern part of South America are a fundamental force governing the horizontal and vertical circulation of the reservoirs of the Paraná basin (Tundisi and Matsumura-Tundisi, 2010; Nyamien and Tundisi, 2017; Nyamien et al., 2019). The mixing that takes place during and after the passage of cold fronts has impacts on the succession of phytoplankton and zooplankton communities of the reservoirs and on biogeochemical cycles (Tundisi et al., 2004).

All the reservoirs of the basin are relatively shallow and during the cold front period they are subjected to winds with velocities of 10 to 20 m s^{-1}, which is sufficient to mix the water column vertically. In studies of two reservoirs in the Tietê River basin (i.e. the Lobo/Broa and Barra Bonita reservoirs), Nyamien and Tundisi (2017) and Nyamien et al. (2019) demonstrated that the circulation dynamics were related to the density profiles in the water column, the force and direction of the wind and the thickness of the transient thermocline. All physical and chemical variables such as conductivity, pH, dissolved oxygen and temperature are distributed relatively homogeneously in the water column during the cold front period. High saturation of dissolved oxygen in the water forces phosphorus into the sediment by precipitation as ferric phosphate. This is one of the factors that turns the reservoirs into ecosystems with high phosphorus retention, which in turn, influences the nutrient budgets within the basin. After the impact of a cold front, a stability period occurs and cyanobacteria (blue green algae) commonly predominate in all eutrophic reservoirs.

However, during the intense period of mixing, chlorophyceae (green algae) and diatoms predominate in the reservoirs. Therefore, the pattern of succession of phytoplankton and zooplankton is established by the frequency and succession of the cold fronts and the stability periods in the reservoirs (Tundisi and Matsumura-Tundisi, 2010). This pulse effect is common to all reservoirs of the basin (Chalar and Tundisi, 1999). An additional variable in this pulse system is the variation of retention time in the reservoirs. The average retention time in the summer months (December to March) is 30 days, whereas the average retention time in winter months (June to September) is 50 days (Tundisi et al., 2008a). Straskraba (1990) observed a relationship between the retention time of reservoirs of the Tietê basin and the phosphorus retention, such that, the higher the retention time, the higher the phosphorus retention, with up to 60% phosphorus retention predicted for reservoirs with retention times of 150 days. Denitrification also plays a very important role in the reservoirs of the basin (Abe et al., 2002). Cold fronts contribute towards the deterioration of water quality of reservoirs that are used for public water supply. Turbidity increases during cold front periods (Tundisi and Matsumura-Tundisi, 2010). Increases in the amounts of suspended sediments in sources of drinking water as a result of re-suspension of sediments is a frequent problem that increases the costs of drinking water treatment. Therefore, water quality monitoring and limnological data are fundamental sources of information for managing drinking water reservoirs. Modelling approaches for water resource management to simulate the effects of cold fronts would improve predictive capacity (Tundisi, 2018)

Eutrophication

The phenomenon of eutrophication occurs in several reservoirs. The Tietê River reservoirs are the most affected because they receive large inputs of organic matter originating from the discharge of untreated wastewater, agricultural fertilizers and industrial effluents. Less than 40% of the domestic wastewater discharged into the river is treated, so the total load of organic matter entering the basin is very high. Another source of nutrients is the non-point contribution of fertilizers from agriculture and the use of feeds with a high phosphorus concentration in aquaculture. For instance, intensive eutrophication was observed in the Lobo/Broa reservoir in 2014 from the previous oligo-mesotrophic trophic status (Tundisi et al., 2015b). Conductivity values for the waters of eutrophic reservoirs in this reservoir are approximately 500 $\mu S\ cm^{-1}$, which is much higher than the conductivity values of 50 $\mu S\ cm^{-1}$ typically observed in oligotrophic reservoirs in the basin (Tundisi and Matsumura-Tundisi, 1990)

Eutrophication results in the growth of cyanobacteria that produce toxins, and because of the degraded water quality, it cannot be used as a water supply for public use, and is not suitable for fisheries, tourism and recreation. In eutrophic reservoirs, the level of total chlorophyll from phytoplankton is approximately 120 $mg\ m^{-3}$, compared to total chlorophyll in oligotrophic reservoirs in the basin of 10 $mg\ m^{-3}$. The most common species of cyanobacteria in the eutrophic reservoirs is *Microcystis aeruginosa*. However, in the last few years, *Cylindrospermopsis raciborski*, an invasive cyanobacteria species which is very toxic, has become more common (Guedes et al., 2014). Figure 4.3

Figure 4.3 A sample of plankton collected during a period of eutrophication in the Lobo/Broa reservoir. This is a bloom of the toxic blue-green algae, *Cylindrospermopsis raciborskii*. Photograph by J. G. Tundisi.

Figure 4.4 *Pistia stratioides* and *Paspalum* spp. macrophytic plants growing in the Lobo/Broa reservoir during a period of intense eutrophication. Photograph by J. G. Tundisi.

illustrates a sample containing almost 99% of this invasive species of cyanobacteria that were present during a period of intense eutrophication in a reservoir. Intensive growth of macrophytic plants, represented mainly by *Eichhornia crassipes* and *Pystiastratioides* also occurs in the eutrophic reservoirs, which interferes with navigation (Figure 4.4).

Transport and accumulation of sediments

Deforestation in the watershed and the intensive use of soil for agriculture are other factors that extensively affect the reservoirs by increasing sedimentation and reducing water quality. There is a high annual loss of top soil in the watersheds of up to 20 tonnes hectare^{-1} (Campagnoli, 2006). The results of high concentrations of suspended sediment in reservoirs are illustrated in Figure 4.5.

However, there is extensive deposition of suspended material into the sediments of reservoirs. For some reservoirs in the Tiete River, sedimentation results in a reduction of 15% of the total volume over a period of 36 years. For instance, the Tres Irmãos reservoir in the Tietê River accumulated sediments to a depth of 18 m over 36 years (Albertin et al., 2010). The sediments cause a range of effects in the reservoirs, including pressure on the dam, damage to pipes, abrasion on the components of the turbines, formation of sand bars that make navigation difficult, loss of spawning areas and feeding habitat for fish, degradation of water quality, high biological oxygen demand (BOD), barriers to light penetration that reduces primary productivity of phytoplankton, and accumulation of heavy metals and toxic substances in the sediments (Albertin et al., 2010). All these impacts increase the costs for maintenance of the reservoir structure and the navigation system.

Figure 4.5 Suspended sediment loads in a reservoir impacted by intensive agriculture along the Tietê River. Photograph by J.G. Tundisi.

Emission of greenhouse gases

The eutrophic reservoirs are also net producers of greenhouse gas emissions to the atmosphere. The accumulation of organic matter in the sediments due to precipitation of particles such as dead organisms and fecal pellets, as well as contributions of organic carbon from the watersheds, followed by metabolism of this material by anaerobic microorganisms in the hypoxic zone is the cause of emissions of a range of greenhouse gases, including methane (CH_4), carbon dioxide (CO_2) and nitrous oxide (N_2O), as described by Straskraba and Tundisi (1999). In a study conducted in three reservoirs of the Tietê River (i.e. Barra Bonita, Ibitinga and Promissão) with a decreasing degree of eutrophication from the upstream reservoir at Barra Bonita to the two downstream reservoirs, Abe et al. (2009) observed a decreasing trend in the emissions of CH_4 and N_2O along the chain of reservoirs. The same pattern was observed for total nitrogen (TN), total phosphorus (TP), nitrate (NO_3) and dissolved organic carbon (DOC). In the rainy season, the concentrations of CH_4, CO_2 and N_2O in the sediments were higher than in the dry season. The higher gas concentration in the rainy season is probably due to the input of allocthonous organic matter from the watershed to the reservoir. The highest concentrations of CH_4, CO_2 and N_2O in the sediments were observed near the dam (Abe et al., 2002, 2009). The higher N_2O emissions in the most eutrophic reservoir was related to the high concentrations of nitrate and nitrite in the water (i.e. precursors of denitrification). Barra Bonita, the most eutrophic reservoir acted as a CO_2 source, while Ibitinga and Promissão, the downstream reservoirs, acted as a CO_2 sink, both in the dry season and in the rainy season (Abe et al., 2009). Therefore, poor management of the water resources and the discharges of untreated sewage into the reservoirs result in high greenhouse gas emissions to the atmosphere. The daily rates of carbon sedimentation measured were near 1,000 mg m^{-2}. Degassing in the outflows of the reservoirs also contribute to the emissions.

There is a relationship between the hydraulic retention times of reservoirs and gas emissions (FURNAS, 2008). In reservoirs with lower retention times (< 20 days), emissions are much lower than in reservoirs with higher retention times (>100 days), which is due to the accumulation of organic matter in the reservoirs with longer retention times. Also, the age of the reservoir is an important factor, as older reservoirs (i.e. >20 years) produce less emissions (FURNAS, 2008). The greenhouse gas emissions for Amazonian reservoirs are much higher than the emissions from the Paraná basin reservoirs. For example, in the Balbina reservoir in the Amazon, daily emission of N_2O is 3.39 mg m^{-2}, while for the Itaipu reservoir in the Paraná basin, the daily emission is 0.16 mg m^{-2} (Projeto Balcar, 2014).

However, in general, the emissions of greenhouse gases by reservoirs of the Paraná basin are lower than the emissions of greenhouse gases by thermoelectric power plants. For example, the maximum value of CO_2 emissions per kw h^{-1} for reservoirs is 91 g, while for coal and gas thermoelectric plants the value is 930 g of CO_2. Table 4.2 shows the theoretical daily diffusive fluxes (mg m^{-2}) across the sediment water interface at different sites within the Estreito reservoir of the Rio Grande River, as reported by Sidagis-Galli et al. (2009).

Impacts of reservoirs

The damming of a river results in a rapid and intense transformation from a lotic to a lentic environment, thus creating new ecosystems with novel structure and function. The main impact of dam construction in a river is the abrupt change in the water flows. Since this is a fundamental force in river ecosystems, the water flow is the dominant force in the geographical/hydrological structure, affecting the dynamics of production of organic matter, nutrient cycling and the distribution and evolution of biota (Baxter, 1977; Poff et al., 1977). The impact is even greater when a succession of reservoirs is placed in series in a river, increasing the fragmentation of habitats and decreasing extensively the water flows. What has occurred in the Paraná River Basin is the construction over the years of a cascade of reservoirs. This unique feature of the basin increases the complexity of the impacts and the management initiatives. According to Petts (1984), impacts from the construction of reservoirs include: ***first order*** impacts that cause physical, chemical, hydrological and geomorphologic changes, ***second order*** impacts that change the productivity of biota and the structure and dynamics of food chains, and ***third order*** impacts

Table 4.2 Daily diffusive fluxes (mg m^{-2}) of methane (CH_4) and carbon dioxide (CO_2) across the sediment-water interface calculated from the mean of three measurements *in situ* from the integration of cores of 0 to 4 cm long collected at upstream to downstream sites in the Estreito reservoir on the Rio Grande River in the Upper Paraná basin.

Sampling site	*Daily diffusive flux (mg m^{-2})*	
	CH_4	*CO_2*
A (upstream)	42.1	1,275
B	53.4	307
C	3.2	254
D	7.9	2.2
E (downstream)	2.5	7.8

that produce major modifications to biological communities, such as fish and invertebrates due to habitat fragmentation.

The new artificial reservoir ecosystems are unstable, and further uses of the watershed, including dam operation and agricultural development, produce further impacts in the reservoir. Table 4.3 describes the main positive and negative impacts of reservoirs, which have occurred in all the reservoirs of the Upper Paraná basin. Although the impacts described in this table could apply to any reservoir that has been constructed in large river basins, there are specific features of the reservoirs that are relevant to the Paraná River Basin that are identified in this table.

The positive impacts of the several reservoirs in the upper Paraná Basin are related mainly to improvements in the economy and water security, while negative impacts include harm to human riverine populations, losses of ecosystem services from the terrestrial and aquatic environments, and reduced water quality and loss of biodiversity. The adaptation of a river culture to a reservoir culture took 30 years in the upper basin (Henry, 2014). Quantitative analysis of the positive benefits from dam construction in the Paraná basin have been studied by several authors (Agostinho et al., 2007a; Tundisi and Matsumura-Tundisi, 2013; Henry, 2014; Tundisi, 2018). The quantification of services and their economic evaluation has been studied in four reservoirs of the basin (Tundisi and Straskraba, 1999; Agostinho et al., 2007a,

Table 4.3 Positive and negative impacts of reservoirs. The impacts that are particularly relevant to reservoirs within the Paraná River Basin are identified with an asterisk (⋆).

POSITIVE IMPACTS
Energy production⋆
Water retention and improvements to water security⋆
Water for public supply⋆
Enhancing local and regional economy⋆
Flood protection
Increase in water supply for irrigation⋆
Introduction of aquaculture⋆
Navigation⋆
Water supply for dry season scarcity
NEGATIVE IMPACTS
River fragmentation and reduction of water flows⋆
Inundation of settlements and human population relocation
Changes in the culture of riverine human population⋆
Spread of waterborne diseases
Loss of native fish species of high commercial value ⋆
Loss of fertile arable land and valuable timber⋆
Habitat alteration for terrestrial and aquatic organisms⋆
Degradation of water quality⋆
Impaired fish migration and spawning of native fish species⋆
Loss of cultural and historical sites for Indigenous populations
Loss of aesthetic value
Reduction of water flows downstream
Loss of wetlands⋆
Introduction of exotic species⋆

2007b; Periotto and Tundisi, 2013). Positive services identified were energy production, flood control, water for irrigation, water reserves and water security for the development of aquaculture, fisheries, tourism and recreation, and transportation of goods and people. For example, for the Carlos Botelho (Lobo-Broa) reservoir, the annual total value of services was estimated as USD $46,000 hectare^{-1} (Periotto and Tundisi, 2013), which is much higher than the services calculated for the period before reservoir construction.

The introduction of exotic species to the Paraná River Basin requires special attention as a negative impact of dam construction. Introduced fish species could occupy empty niches in the reservoirs such as the pelagic zone and produce several changes in the food chain. At present, 32 species of fishes are cultivated in aquaculture facilities in the watersheds around the reservoirs or in aquaculture cages within the reservoir itself. Despite the benefits of high production of protein, aquaculture is a source of eutrophication from inputs of food and feces, and a potential source for the accidental release of exotic species (Smith et al., 2005). Mollusk species that have been introduced to the basin as a result of navigation by ocean going ships include the gastropod, *Melanoides tuberculate* and the bivalve, *Corbicula fulminea* (Pamplin and Rocha, 2005). *Limnoperna fortune* is another introduced bivalve that has had a great impact (Pareschi et al., 2008). Direct competition with native species in the shallow areas of the reservoir produced several changes in the populations of native mollusk species. The introduced species can also block pipes and interfere with the infrastructure of hydroelectric power plants. Human health is also affected, as *Melanoides tuberculate* is an intermediate host of a trematode parasite and *Corbicula fulminea* is a host and transmitter of intestinal diseases caused by *Cryptosporidium parvum* and *Giardia lamblia* (Graczyk et al., 2003).

Management strategies

Given the scale of the basin, a regional approach for management of watersheds-reservoirs seems to be appropriate. The ecological dynamics of the reservoirs depends upon the relationship with the watershed, its operational requirements and the multiple uses for the impounded water. Since the reservoirs of the Paraná River Basin have been extensively studied over the last 30 years (Agostinho et al., 2008; Tundisi and Matsumura-Tundisi, 2013; Tundisi, 2018), a framework for the management of these artificial ecosystems has emerged from the accumulated knowledge of their basic ecological functioning, impacts and responses. This framework includes the following initiatives:

- **Institutional organization** to address the sustainability of the Basin. For instance, the Government of São Paulo State created 21 watershed committees in order to decentralize the management and the financing of the sustainability program.
- **Participation** of all stakeholders in the committees in order to assure representation of all water users.
- **Technologies** that include physical interventions such as reforestation of riparian forests, improved wastewater treatment, control of erosion and soil degradation and scientific information using new real time monitoring techniques, programs for monitoring water quality and water protection of source areas for water supply.
- **Knowledge dissemination** to shape public perception about the ecological status of the reservoirs and the management strategies. For some reservoirs, such as Carlos Botelho (Lobo/Broa, Tietê watershed)), Itaipu Reservoir (Paraná River), Jurumirim (Paranapanema River), education of the population on environmental issues, capacity building for technical

staff, intensive monitoring and social actions proved to be a useful approach for promoting the management of the local watershed and reservoirs (Tundisi, 1987). New and advanced technologies for translating scientific information for dissemination to the general public were effective. Combined with modern and improved local and regional legislation and increasing public awareness, this initiative was effective in promoting the protection and recovery of watersheds and reservoirs (Tundisi and Matsumura-Tundisi, 2013). Protection of wetlands, special management programs for fisheries and aquaculture were implemented for reservoirs in the Tietê and Paraná rivers (Agostinho et al., 1999).

- **Financing** to provide local or regional support for implementing management and recovery initiatives. The Government of São Paulo State in Brazil has decentralized the application of financial support for research and management of the watersheds and reservoirs. The 21 watershed committees created all over São Paulo State based on the Brazilian Water Law of 1997 receive the funds from penalties assessed for pollution and deforestation (Tundisi et al., 2008a). Additional funding sources include the taxes from polluter–pay schemes and from payments for environmental services. For example, the Tietê-Jacare watershed, which includes 34 municipalities and 1.2 million inhabitants in an area of 3.2 million km^2 with two reservoirs, applied in 2018 for the equivalent of USD \$1 million to support research and management activities.

The management of watersheds and reservoirs should also include consideration of climate change impacts (Pittock, 2010). Climatic changes could affect the reservoirs by reducing precipitation and the water supply, by increasing water temperatures that promote the growth of cyanobacteria, and by affecting geochemical cycles. An example of these effects was described by Tundisi et al. (2015b) for the Carlos Botelho (Lobo/Broa) reservoir in the upper basin of the Paraná River during a period of a large-scale climatic change in 2013 and 2014. The consequence of this short-term climatic crisis was a loss of USD \$10 million in the Hidrovia Tietê/Paraná waterway due to the shutdown of its transportation operations for 10 months (Tundisi et al., 2015a).

An overall perspective for the management of the reservoirs is the concept of an ecohydrological approach that integrates hydropower production, development and sustainability (Tundisi and Matsumura-Tundisi, 2016). Zalewski (2014) argued that the integration of management schemes to protect water resources, biodiversity, ecosystem services, resilience and cultural heritage is an optimal approach. The integration of engineering, ecological principles and hydrology is becoming an important management perspective in the Paraná River Basin. Several small reservoirs in the Upper Paraná River sub-basin have established management projects that are similar to this model, proving that a regional watershed approach can be very effective. A watershed-based approach has been proposed as a fundamental action plan for basin management (Likens, 1992).

An integrated management plan was developed for the Lobo/Broa reservoir and the Itaqueri River as part of a larger plan for the management of the Tietê-Jacaré basin. The elements of the plan are to:

1. **Develop basin water quality standards**
   - Irrigation water quality standard
   - Drinking water quality standard
   - River water quality standard
   - Sewage effluent quality standard.

2 **Develop basin water quality regulations and plans**
- Agrochemical regulations
- Total maximum water withdrawals
- Ground water extraction regulations
- Mining regulations
- Dairy waste reuse and management plan
- River environment management plan.

3 **Develop basin infrastructure plans**
- River basin management plan
- Regional sewage development plan
- Land use and zoning plan
- Protection and recovery of gallery forests and vegetation mosaics
- Water resource management plan.

The management plan for this river and reservoir aims to: (i) introduce regulations to control pollution discharges into the watershed and the reservoir, (ii) carry out intensive and comprehensive reservoir water improvement measures and develop active cooperation between local governments, municipalities that share the watershed and the reservoir (Nakamura and Rast, 2011). Tundisi et al. (2018) argued that close institutional cooperation, community participation and intensive monitoring is essential for completing this task.

Summary

The large-scale construction of artificial ecosystems (i.e. reservoirs) and the intensive investment in associated infrastructure, although very important for water security and economic development, changed irreversibly the natural ecosystems of the Upper Paraná basin, resulting in negative impacts within the entire watershed, its terrestrial and aquatic ecosystems and the human river culture. On the other hand, the multiple uses of the reservoirs promoted benefits for local and regional populations, in terms of economic development, employment opportunities and industrial and agricultural production. Large investments in infrastructure and regional modernization occurred. The water infrastructure projects attracted migrations of human populations, enhanced agricultural activities and expanded industrial production. However, intensive use of soils, discharges of untreated wastewater and industrial effluents, deterioration of water quality and reduced ecosystem services followed (Tundisi et al., 2015a). Therefore, the reservoirs in the Upper Paraná River Basin provide a classic example of modernization that first enhances the economy and diversifies multiple uses of natural resources, but later negatively affects ecosystem services, beneficial water uses and human health, thereby eventually impairing sustainable development. Modernization and development are quite different processes.

Even though Environmental Impact Assessments (EIA) were conducted prior to construction of many of the hydropower dams, the EIAs did not provide perspectives on sustainable development of the watersheds and their reservoirs. Frequently, EIAs fail to influence decision making among stakeholders. The environmental risks of dam projects and associated economic investment were not incorporated as key factors in the decision-making process, as they were not in many other watersheds of the world (World Commission on Dams, 2000). Strategic environmental planning and management approaches were not considered simultaneously with the development of large-scale water infrastructure, reservoir construction and the implementation of multiple uses. The investments in planning and incorporation of measures for environmental management and protection would have been a relatively small fraction of the total economic input required at the beginning of the process. At present, the retrospective costs of actions to promote the sustainability of the basin are much greater (Milikan, 2010).

A strategy of dividing the basin into regional units for managing watersheds-reservoirs should be an appropriate approach for such a large-scale system with great regional-latitudinal differences in economy, cultures, water uses and ecological awareness among populations. Ecological approaches that integrate environmental management with ecology as an approach for protecting water resources, biodiversity, ecosystem services, resilience and cultural heritage are essential for this complex system (Tundisi, 2018; Matsumura-Tundisi and Tundisi, 2018). New and creative financial, institutional and coordination initiatives are in progress at several reservoirs and watersheds within the basin, and therefore collaboration between the number of agencies operating within the basin is urgently needed.

References

Abe, D.S., Rocha, O., Matsumura-Tundisi, T., Tundisi, J.G. 2002. Nitrification and denitrification in a series of reservoirs in the Tietê River, Southeastern Brazil. *Verhandlungen des Internationalen Verein Limnologie* 28:877–880.

Abe, D.S., Sidagi-Galli, C., Matsumura-Tundisi, T., Tundisi, J.E.M., Grimberg, D., Medeiros, G., Teixeira-Silva, V., Tundisi, J.G. 2009. The effect of eutrophication greenhouse gas emissions in three reservoirs of the middle Tietê River, southeastern Brazil. *Verhandlungen des Internationalen Verein Limnologie* 30:822–825.

Agostinho, A.A., Gomes, L.C., Pelicice, F.M. 1999. *Ecologia e manejo de recursos pesqueiros em reservatórios do Brasil*. Editora Universidade Estadual de Maringá, Brazil, p. 501.

Agostinho, A.A., Gomes, L.C., Pelicice, F.M. 2007b. *Ecologia e manejo de recursos pesqueiros em reservatórios do Brasil*. EDUEM, Maringá University, p. 501.

Agostinho, A.A., Pelicice, F.M., Gomes, L.C. 2008. Dams and fish fauna of the neotropical region: Impacts and management related to diversity and fisheries. *Brazilian Journal of Biology* 68(suppl 4):1119–1152.

Agostinho, A.A., Pelicice, F.M., Petry, A.C., Gomes, L.C., Julio, H.F., Jr. 2007a. Fish diversity in the upper Parana River basin: Habitats, fisheries management and conservation. *Journal of Aquatic Ecology and Health Management* 10:174–186.

Albertin, L.L., Matos, J.S.A., Mauad, F.F. 2010. Cálculo do volume e análise de deposição de sedimentos do reservatorio Três Irmãos. *Revista Brasiliera de Recursos Hidricos* 54:57–67.

Baxter, M. 1977. Environmental effects of dams and impoundments. *Annual Reviews in Ecological Systems* 8:255–285.

Campagnoli, F. 2006. Production of sediments from South America: A proposal of erosion rates mapping based on geological and geomorphological data characteristics. *Revista Brasileira de Geomorfologia* 7:3–8.

Chalar, G., Tundisi, J.G. 1999. Main processes in the water column determined by wind and rainfall at Lobo (Bro) reservoir. Implications for phosphorus cycling. In: Tundisi, J.G., Straskraba, M. (eds.), *Theoretical Reservoir Ecology and Its Applications*, Backhuys Publishers, Rio de Janeiro, Brazil, pp. 53–66.

FURNAS. 2008. *Projeto de P & D, Aneel. Balanço de carbono nos reservatórios de Furnas Centrais Elétricas S.A., Relatório Final*. Furnas Centrais Elétricas S.A., Rio de Janeiro, Brazil.

Graczyk, T.K., Conn, D., Marcogliese, D., Graczyk, H., De Lafontaine, Y. 2003. Accumulation of human waterborne parasites by zebra mussels (*Dreissena polymorpha*) and Asian freshwater clams (*Corbicula fluminea). Parasitology Research* 89:107–112.

Grananda, D.C., Romero, L.R. 2014. Lazer e turismo na represa de Jurumirim. In: Henry, R. (ed.) *A Reprea de Jurumiram: Ecologia, Modelagem e Aspectos Sociais*. Holos Editora, Ribeirão, Brazil, pp. 367–376.

Guedes, I.A., Costata Leite, D.M., Manhães, L.A., Brisch, P.M. 2014. Fluctuations in microcystin concentrations, potentially toxic microcystis and genotype diversity in a cyanobacterial community from a tropical reservoir. *Harmful Algae* 39:303–309.

Henry, R. (Ed.) 2014. *Represa de Jurumirim: Ecologia, Modelagem e Aspectos Sociais*. Holos Editora, Ribeirão, Brazil, p. 434.

Kennedy, R.H. 1999. Reservoirs design and operation: Limnological implications and management opportunities. In: Tundisi, J.G., Straskraba, M. (eds.), *Theoretical Reservoir Ecology and Its Applications*, International Institute of Ecology, Brazilian Academy of Sciences, Backhuys Publications, Rio de Janeiro, Brazil, pp. 1–28.

Likens, G.E. 1992. *The Ecosystem Approach: Its Use and Abuse.* Excellence in Ecology Series, Ecology Institute, Oldendorf, Germany, p. 166.

Matsumura-Tundisi, T., Tundisi, J.G. (eds.) 2018. *Water Resources Management.* Brazilian Academy of Sciences, Inter-American Network of Academies of Sciences, UNESCO Regional Bureau for Science in Latin America and the Caribbean, Editora Scienza, Sao Carlos, Brazil, p. 248.

Milikan, B. 2010. Brazil's dam building industry. *Crony capitalism goes global. World Rivers Review* 25:4–45.

Nakamura, M., Rast, W. 2011. *Development of ILBM Platform Processes: Evolving Guidelines through Participatory Improvement.* RCSE-SU, ILEC, Japan, 76 p.

Nyamien, Y.S., Tundisi, J.G. 2017. *Repercussão da climatologia e hidrologia sobre o reservatorio.* Estudo de caso: reservatório de Barra Bonita, SP; Carlos Botelho, SP e Tucurui, PA, Novas Edições Academicas, Scarbruchen, Deutschland/Niemcy, p. 156.

Nyamien, Y.S., Tundisi, J.G., Batista, L.C., Nogueira, D.V. 2019. Dynamics of non- dimensional, Richardson, Wedderburn and lake numbers applied to neotropical reservoirs Barra Bonita and Carlos Botelho (Broa) São Paulo, Brazil. *International Journal of Applied Science and Technology* 9:67–78.

Pamplin, P.A.Z., Rocha, O. 2005. Distribuição altimétrica e sazonal de moluscos na represa de Bariri, com ênfase nas espécies exóticas, *Melanoidestuberculata* (Gastropoda: Thidaridae) e *Corbicula fluminea* (Bivalva: Corbiculi). In: Odete, R., Espindola, E.L.G., Fenerich-Verani, N., Verani, J.R., Rietzler, A.C., (organizers), *Espécies invasoras em aguas doces: Estudos de caso e propostas de manejo*, Editora Universidade Federal de São Carlos, Brazil, 416 p.

Pareschi, D.C., Matsumura-Tundisi, T., Medeiros, G.R., Luzia, A.P., Tundisi, J.G. 2008. First occurrence of *Limnoperna fortunae* (Dunker 1857) in the Rio Tietê watershed-São Paulo State, Brazil. *Brazilian Journal of Biology* 68(suppl 4):1107–1114.

Periotto, N., Tundisi, J.G. 2013. Ecosystem services of UHE Carlos Botelho (Lobo/Broa reservoir): A new approach for the management and planning of dams multiple uses. *Brazilian Journal of Biology* 73:471–482.

Petts, G.E. 1984. *Impounded Rivers: Perspectives for Ecological Management.* John Wiley & Sons, Chichester, p. 326.

Pittock, J. 2010. Better management of hydropower in an era of climate change. *Water Alternatives* 3:144–452.

Poff, N.L., Allan, J.D., Bain, M.B., Kan, Jr, Prestegcard, K.L., Richter, B.D., Spanks, R.E., Strumberg, J.G. 1977. The natural flow regime. *Bioscience* 47:769–784.

Projeto Balcar. 2014. *Emissões de gase de efeito estufa em reservatórios de centrais hidrelétricas.* ANEEL, Eletrobrás, Ministerio de Minas e Energía, Brasilia, Brazil, p. 416.

Rocha, O., Espindola, E.L.G., Fenerich-Verani, N., Verani, J.R., Rietzler, A.C. 2005. *Espécies invasoras em águas doces. Estudos de caso e propostas de manejo.* Editora Universidade Federal de São Carlos, Brazil, p. 417.

Sidagis-Galli, C., Abe, D.S., Tundisi, J.G., Teixeira-Silva, V., Medeiros, G.R.; Brum de Barros, P.R.H., Cimbleris, A.C.P., 2009. Greenhouse gas concentrations and diffusive flux at the sediment-water interface from two reservoirs in Brazil. *Verhandlungen des Internationalen Verein Limnologie* 30(Part 6):830–833.

Smith, W.S., Espindola, E.L., Rocha, O. 2005. A introdução de espécies de peixes exóticos e alóctones em bacias hidrográficas brasileiras. In: Rocha, O., Espindola, E.L.G., Fenerich-Verani, N., Verani, J.R., Rietzler, A.C. (eds.) *Espécies invasoras em águas doces: estudos de caso e propostas de manejo.* Editora Universidade Federal de São Carlos, Brazil, pp. 15–44.

Stevaux, J.G., Martins, D.P., Meuner, M. 2009. Changes in a large regulated tropical river: The Paraná River downstream from the Porto Primavera dam, Brazil. *Geomorphology* 113:230–238.

Straskraba, M. 1999. Retention time as key variable of reservoir limnology. In: Tundisi, J.G., Straskraba, M. (eds.), *Theoretical Reservoir Ecology and Its Applications.* International Institute of Ecology, Brazilian Academy of Sciences, Backhuys Publishers, Rio de Janeiro, Brazil, 385–410 pp.

Straskraba, M., Tundisi, J.G. 1999. Reservoir ecosystem functioning: Theory and application. In: Tundisi, J.G., Straskraba, M. (eds.), *Theoretical Reservoir Ecology and Its Application.* International Institute of Ecology, Brazilian Academy of Sciences, Backhuys Publishers, Rio de Janeiro, Brazil, pp. 565–597.

Tundisi, J.G. 1987. *Local Community Involvement in Environmental Planning and Management: The Lobo/Broa Reservoir Case Study.* Regional Development Dialogue, UN Centre for Regional Development, Nagoya, Japan, pp. 135–155.

Tundisi, J.G. 1994. Tropical South America: Present and perspectives. In: Margalef, R. (ed.) *Limnology Now: A Paradigm of Planetary Problems.* Elsevier, Amsterdam, The Netherlands, pp, 353–424.

Tundisi, J.G. 2018. Reservoirs: New challenges for ecosystem studies and environmental management. *Water Security* 4–5:1–7.

Tundisi, J., Matsumura-Tundisi, T. 1990. The limnology and eutrophication of Barra Bonita reservoir. São Paulo State, Southern Brazil. *Archives of Hydrobiology* 33:663–666.

Tundisi, J.G., Matsumura-Tundisi, T. 2010. Impactos potenciais das alterações do Código Florestal nos recursos hídricos. *Biota Neotropical* 10:67–76.

Tundisi, J.G., Matsumura-Tundisi, T. (eds.) 2012. *Limnology*. Taylor & Francis, London, p. 832.

Tundisi, J.G., Matsumura-Tundisi, T. 2013. The ecology of the UHE Carlos Botelho (Lobo/Broa) reservoir and its watershed, São Paulo, Brazil. *Freshwater Reviews* 6:75–91.

Tundisi, J.G., Matsumura-Tundisi, T. 2016. Integrating ecohydrology, water management and watershed economy: Case studies from Brazil. *Ecohydrology and Hydrobiology* 16:83–91.

Tundisi, J.G., Straskraba, M. 1999 (Editors) *Theoretical reservoir ecology and its applications. International Institute of Ecology*, Brazilian Academy of Sciences, Backhuys Publishers, Rio de Janeiro, Brazil, 585 pp.

Tundisi, J.G., Matsumura-Tundisi, T., Arantes Junior, J.D., Tundisi, J.E.M., Manzini, N.F., Ducrot, R. 2004. The response of Carlos Botelho (Lobo/Broa reservoir to the passage of cold fronts as reflected by physical, chemical and biological variables. *Brazilian Journal of Biology* 64:177–186.

Tundisi, J.G., Matsumura-Tundisi, T., Abe, D.S. 2008a. The ecological dynamics of Barra Bonita (Tietê River, SP, Brazil) reservoir: Implications for its biodiversity. *Brazilian Journal of Biology* 68(4 suppl.):1079–1098.

Tundisi, J.G., Matsumura-Tundisi, T., Pareschi, D.C., Luzia, A.P., Healing, P.A., Von, Frollini, R. 2008b. A bacia hidrográfica do Tietê/Jacaré: Estudo de caso em pesquisa e gerenciamento. *Estudos Avançados* 22:159–172.

Tundisi, J.G., Matsumura-Tundisi, T., Ciminelli, V.S., Barbosa, F.A. 2015a Water availability, water quality, water governance: The future ahead. International Association of Hydrology. *Proceedings of the 11th Kovacs Colloquium*, Paris, France, June, 2014, 366:75–79.

Tundisi, J.G., Matsumura-Tundisi, T., Tundisi, J.M., Blanco, F.P., Abe, D.S., Sidagis Galli, C., Campanelli, L.C., Silva, V.T., Lima, C.P.P. 2015b. A bloom of cyanobacteria (*Cilindrospermopsis raciborski*) in UHE Carlos Botelho (Lobo–Broa) reservoir: A consequence of global change? *Brazilian Journal of Biology* 75:507–508.

Tundisi, J.G., Matsumura-Tundisi, T., Tundisi, J.E.M. 2018. Integrated management plan for the Itaqueri/Lobo watershed and UHE Carlos Botelho (Lobo/Broa) reservoir. In: Matsumura-Tundisi, T., Tundisi, J.G. (eds.) *Water Resources Management*. Editora Scienza, Sao Carlos, Brazil, 248 p p.

Valencio, N., Gonçalves, J.G. 2014. Do alarde desenvolvimentista o silencio inoportuno sobre os riscos: Limites da convivencia social comum a grande barragem, pp 343–366. In: Henry, R. (ed.) *A Represa de Jurumiram: Ecologia, Modelagem e Aspectos Sociais*. Holos Editora, Ribeirão Preto, Brazil, p. 434.

World Commission on Dams. 2000. *Dams and Development. A New Framework for Decision Making*. Earthscan, London.

Zalewski, M. 2014. Ecohydrology, biotechnology and engineering for cost efficiency in reaching the sustainability of the biogeosphere. *Ecohydrology and Hydrobiology* 70:689–695.

5

Ecosystems

Pablo Collins and Mercedes R. Marchese

Key messages

- The Paraná River is characterized by a high spatial and temporal heterogeneity that includes a wide variety of permanent and temporary ecosystems regulated by the hydro-sedimentological regime and hydrological connectivity.
- The longitudinal and lateral connectivity between rivers and floodplains is vital to maintain the functionality of ecosystems within the basin and to maintain its high biodiversity. Development by man threatens the connectivity between aquatic environments, posing a risk to biodiversity.
- In the basin of the Paraná River there are several large wetlands, currently characterized by minimal intervention by man, except in the stretch of Upper Paraná, where there are many hydroelectric dams and reservoirs.

- Unlike many other great river systems around the world, the headwaters of the Paraná River are located in a tropical region and the river passes in a southerly direction through subtropical and then temperate regions, which contributes to changes in biotic diversity from north to south.
- The high biodiversity includes faunal groupings from benthic invertebrates and zooplankton to mammals, and the aquatic plants are essential components of the flora, providing refuge and food for the fauna.
- The complexity of the system, together with the lack of effort by the countries within the basin to investigate the species diversity has contributed to uncertainty regarding the conservation status and risks to biodiversity. Therefore, many species are identified as having Data Deficient (DD) status in the IUCN Red List.

Introduction

Freshwater ecosystems comprise less than 1% of the Earth's surface but contribute disproportionately to Earth's biodiversity, providing habitat for approximately 10% of the Earth's species (Kawai and Cumberlidge, 2016). The elimination of species at any particular time of the seasonal cycles in some areas does not produce catastrophic effects due to compensatory mechanisms of the ecosystem of the Paraná basin, as evidenced by the fact that the current biota assemblage of the Paraguay-Paraná system is several thousand years old (Neiff et al., 2000).

The natural hydrological regime in the Upper Paraná basin has been severely altered by the construction of reservoirs, diminishing the amplitude of seasonal water level fluctuations, decreasing the flood peaks and causing new minimum water levels. There are differences in the ecological characteristics during the phases of the hydro-sedimentological cycle. During low water periods, floodplain habitats are less connected with the river and are subject to local processes like wind turbulence and inputs from small tributaries. During high water levels, species assemblages show greater similarities and are widespread in different habitats.

The Paraná fluvial system forms an ecological corridor that crosses the central part of South America from the tropical headwaters (closely linked with Amazonian head water streams) to the temperate zone of the Río de la Plata estuary, as illustrated in Figure 5.1. This fluvial corridor provides connectivity between landscapes, ecosystems and natural or modified habitats, maintains biological diversity through the genetic exchange of flora and fauna species, and supports biogeochemical flows. The connectivity between ecosystems is only possible when the hydrological cycles are not altered over the entire basin. This great central depression in South America comprises a hydrological and biological continuum of wetlands, which includes, from North to South, the Pantanal of the Upper Paraguay, the marshes of the Chaco, the wetlands of San Pedro, Ypacarai, Ypoa and Ñeembucu in Eastern Paraguay, the wide floodplain of the Paraná River, the Esteros del Iberá, and finally, the Paraná Delta, forming the most extensive wetland system on the planet that is drained by a continuous main channel of almost 4,000 kilometers of rivers that are free of dams (Neiff, 2005).

Ecological functioning of a floodplain system must be viewed in a multidimensional manner, incorporating lateral, longitudinal and vertical changes, as well as changes over time. Within this multidimensional template, transport of organic matter, energy flows and genetic material have fundamental ecological importance for the functioning of the fluvial system. Rivers and their floodplains form important corridors for the migration of

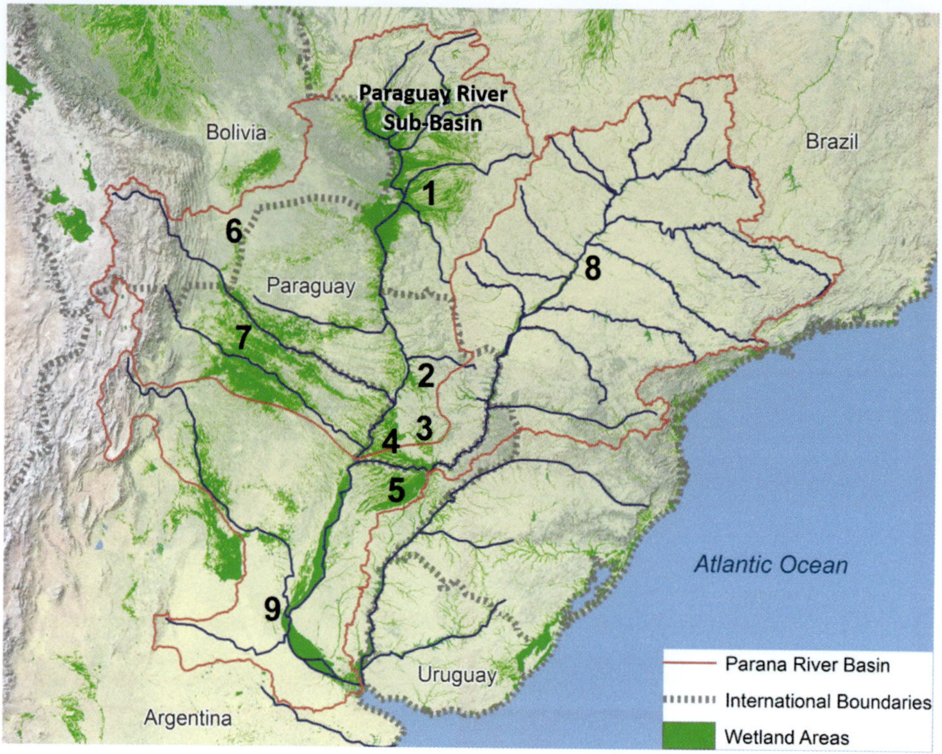

Figure 5.1 Primary wetland areas throughout the basin of the Paraná River: 1. Pantanal, 2. San Pedro, 3. Ypacarai–Ypoa, 4. Ñeembucu, 5. Iberá, 6. Izozog, 7. Chaco marshes, 8. Multiple reservoirs of hydroelectric dams, 9. Floodplains of the Middle Paraná and the Delta. Hamid Mehmood compiled the geospatial data in ArcMap format and Erin Hayward prepared the map using those data.

organisms both in the downstream and the upstream direction. In addition to the migrations of plants and animals along the river corridor, human populations have also used it as migratory routes.

Major ecosystems in the basin

The hydrography of the Paraná River Basin is dominated by large rivers and associated ecosystems such as extensive floodplains, secondary channels and shallow lakes, whereas there are few large natural lakes. There have been many changes in the river system as a result of human activities, and one of the major impacts has been through the construction of dams that alter the flow conditions (Agostinho et al., 2008). Approximately 140 dams and reservoirs have been constructed in the Upper Paraná River Basin, and the only remaining area of the original flood plain is a 230 km stretch between the upstream margin of the Itaipú Reservoir and the Porto Primavera Dam (Agostinho et al., 2008).

At smaller spatial scales, it is possible to distinguish within each type of ecosystem a great variety of habitats. These habitats are differentiated mainly by specific physical variables, such as depth, current velocity, substrate type, hydraulic conditions, position within the ecosystem (e.g. thalweg and banks in rivers, open water and littoral areas in lakes, tributary confluences), but also by water quality.

Main channel

The Paraná River Basin is represented by a complex mosaic of habitats that are undergoing ecological succession within both regulated and unregulated sections of the river. Controlling factors, such as flood and low-water phases, tributary inputs, riparian influences, groundwater seepage, hydraulic conditions, etc., interact to influence each aquatic habitat.

In the upstream reaches of the Upper Paraná, the protruding basaltic layer forms riffles, rapids and falls, as well as modified landscapes that have been formed by the hydroelectric dams. Then, downstream of the mouth of the Iguazú River, the Paraná River flows in a single sinuous channel deeply cut into the basaltic substrate, flowing on a rocky bed with large sand patches up to the west side of the city of Posadas in Argentina. Downstream of the confluence with the Paraguay River, the channel bed of the Paraná River is increasingly covered by sand rather than rock, and channel bars and islands have been formed (Iriondo et al., 2007). The downstream portion of this system is complex, with numerous low islands and an extensive floodplain which progressively widens along the western bank (Bonetto and Wais, 1995). The region consists of a complex ensemble of lotic environments of diverse size (Figure 5.2, photos 1 and 2), with variable discharge and temporal dynamics and there are also many low islands that formed during different episodes of accumulation and sedimentation. In this floodplain system, varied lentic environments also occur, which are coupled to the drainage network (Figure 5.2, photos 3 and 4). Some of these lentic environments are dry during low water seasons, but others are more stable and permanent.

The Paraguay River, with a length of nearly 2,800 km, drains a sub-basin that extends into southwestern Brazil and also into the foothills of the Andes mountains in Bolivia. In the Upper Paraguay, several rock outcrops of quartz and feldspar sandstones emerge in the main channel and the slope varies between 0.07 and 0.65 m km^{-1}, with an average value along the entire Paraguay River of 0.04 m km^{-1}. Due to a weak slope, the Pantanal wetland delays the flow of water, nutrients and sediments into the river. The prevailing climate in the Upper Paraguay catchment varies from dry to sub-humid, with mean annual rainfall of about 1,180 mm, 85% of which falls during the rainy season (Wantzen et al., 2005). The low mean concentrations of suspended sediment measured in Asunción city (i.e. 68 mg L^{-1}) reflect the capacity of the Pantanal wetland in the Upper Paraguay basin to trap sediments (Drago et al., 2008). The Lower Paraguay River is characterized by a meandering channel and extensive floodplains. It has three important tributaries: the Bermejo, Pilcomayo and Tebicuary Rivers. The Pilcomayo and Bermejo rivers, which have their sources in the Andean foothills, have a strong influence on the Paraguay River by carrying large amounts of suspended solids (Amsler and Drago, 2009). The load of solids from the Bermejo River strongly influences the suspended load of the Lower Paraguay. Thus, the high suspended sediment concentration measured in the Bermejo River (i.e. annual average of 6.5 g L^{-1}) is responsible for the sharp increase in the suspended sediment concentration of the Paraguay River from 49 to 313 mg L^{-1} above and below the confluence with the Bermejo, respectively (Drago et al., 2008; Blettler et al., 2014). The concentrations obtained in other large rivers were lower than those in Lower Paraguay, thus in the Uruguay (36 mg L^{-1} mg/L), Madeira (246 mg L^{-1}), Amazonas (99 mg L^{-1}) and Solimoes (230 mg L^{-1}), as reported by Lima et al. (2005). The average concentration of suspended sediment during the period of maximum discharge, for the decade 1990–1999 was calculated at 33 mg L^{-1} for the Upper Paraná at its confluence with the Paraguay River (Amsler and Drago, 2009).

Figure 5.2 A Google Earth image of the Paraná River in the middle stretch showing the types of representative ecosystems, and the extent of urbanization within the cities of Santa Fe and Paraná. Photo 1: Main channel of Paraná River. Photo 2: Secondary channel. Photo 3: Floodplain connected lake. Photo 4: Floodplain isolated lake. Photo 5: Fluvial marginal wetland during high water level.

The main channel of Paraná downstream of the confluence with the Paraguay River has a braided pattern with a fringing floodplain, showing a sequence of wide reaches with two or more anabranches and narrow and short reaches in which the river flows in a single channel. The average main channel width is 2.8 km (range between 0.4 to 8 km), with a maximum depth of 45 m. The braided reaches are the most geomorphologically unstable riverine habitats of the Paraná watershed (Drago, 1990). The riverbed of the Paraná River at this point is predominantly mobile, with many bars and islands, as described in Chapter 3 in this book. Bed sediments almost uniformly consist of sand (mean percentages always over 98%), composed mainly of medium sands that are transported principally by saltation (Drago and Amsler, 1998). The cross-sections of the main channel show two clearly different areas, the central strip and bank strips along the riverbank. The former occupies 90–95% of the wetted perimeter, being the more homogeneous area of the river, with mobile sandy bed materials and limited physical changes over the seasons or with changes in water level.

In the Lower Paraná, the geomorphological characteristics of a sandy bed are maintained, but the granulometry shows a shift to finer sands. From the point of view of the substratum, it is possible to define a longitudinally homogeneous fluvial habitat of 1,447 km in length, where the cross-sections along the main channel present the same pattern, with no less than 95% of wetted perimeter composed of sand (Marchese et al., 2002). Bedforms range from ripples and dunes to sand waves. During high-water levels, the dunes move downriver at a mean velocity of 12 m day^{-1} and the average rate of movement of superimposed ripples reaches 37 m day^{-1}. Whereas, during low water levels the rate of movement decreases to 3.3 m day^{-1} for dunes, and 9.9 m day^{-1} for superimposed ripples (Amsler and Gaudin, 1994). The current velocity in the main channel is always high, ranging from 0.5 to 2.6 m s^{-1} under low to moderate discharges and often exceeding 3 m s^{-1} during high flows.

From latitude 32°05′S to 34°29′S, the Paraná River has developed a fluvio-deltaic complex, which geomorphologically is divided into two sub-regions, known as the upper and the lower delta. The upper delta lies between Diamante city and the diversion point between Paraná Guazú and Paraná de las Palmas. The lower delta extends between this point and the discharge into the Río de la Plata estuary and the delta system grows by an average of 58 m year^{-1} (Drago and Amsler, 1998). The Paraná Guazú, is the most important deltaic tributary, as it presents a higher hydraulic gradient that allows this part of the river to capture 79% of the discharge and most of its sediment load, which at this point is approximately six times greater than that of the Paraná de las Palmas deltaic tributary. Consequently, Paraná Guazú represents the morphologic and sedimentological continuation of the Paraná River into the delta (Drago and Amsler, 1998). It presents predominantly as a sandy bed (99% of total bed load) up to the Río de la Plata estuary.

Floodplain systems

A distinct feature of floodplain systems is high spatial and temporal heterogeneity, which is associated with low predictability within the hydrological system. Such spatial heterogeneity is characterized by the presence of mosaics of micro- and macro-habitats, or distinct geomorphological units, soil and vegetation types, often with different ecological boundaries between them (Figure 5.2). Thus, the floodplain habitats support communities that are distinctive in structure, composition and productivity.

The floodplain environments are subject to constant change during the annual water regime that causes pulses of floods and droughts. The periods between low water phases are characterized by high ecological productivity. The stages of renewal governed by the summer floods have great ecological significance for river-floodplain systems. The extremes in hydrological events are coupled with the phases of the southern oscillation (Camilloni and Barros, 2003), characterized by floods during El Niño events and droughts during La Niña events that alternate with periods of neutral conditions. These extraordinary events produce changes in the river's physical and chemical conditions (Neiff, 1996) and the entire alluvial plain may be covered by one continuous water mass during exceptional floods (Thomaz et al., 2007). The low predictability of ecological processes in floodplains, especially over shorter timescales, is predominantly correlated with fluctuations in water level and the influence of flood pulses. All organisms occupying a floodplain, including humans, must adapt themselves to this low predictability and inherent instability.

Floodplain channels

The floodplain channels or secondary channels have similar functional composition as the main channel and comprise the drainage network of the complex floodplain. They may have permanent flows with high discharge (>2000 $m^3 s^{-1}$) or the flows may be low (< 500 $m^3 s^{-1}$) or ephemeral. The intermittent nature of temporary channels is linked with annual hydrological behavior of the main channel of the river (Iriondo et al., 2007). In addition, differentiation between central and bank strips is not clear in smaller channels (Iriondo et al., 2007; Wantzen et al., 2008). The lower channels with mean discharges of less than 200 $m^3 s^{-1}$ have bottom sediments with dominant silt-clay composition (Iriondo et al., 2007; Blettler et al., 2019). The main stem banks and floodplain channels are changing continuously as a result of erosion or deposition. These channels can be completely covered by macrophytes during summer, and the vegetation, in turn, changes with the hydraulic conditions that drive the rates of the erosion-deposition in the channels (Marchese et al., 2002; Drago et al., 2003; Iriondo et al., 2007). They may change their flow direction according to river level fluctuations (Mesa et al., 2012). On the other hand, channels divided by islands and bars within the main channel are termed anabranches; that is, they flow parallel to one or both sides of deepest channel (Drago et al., 2003).

Floodplain lakes

The floodplain lakes or 'lagoons' according to some authors, show hydrological heterogeneity because of different degrees of connectivity with channels, varying from permanently connected lakes to those lakes that exchange surface water only during maximum flood peaks (Figure 5.2, Photos 3 and 4). Their surfaces may be free of vegetation or covered totally or partially by emergent macrophytes. The hydraulic residence times vary according to the distance of the lakes from the active channels and the type and degree of connection. Moreover, hydraulic residence times can also vary across microhabitats within a particular lake. The variability of microhabitats within a lake will depend on the origin, morphometrics and evolution of its basin. Thus, it is possible to find different successional stages, according to the degree of connectivity and the location in the floodplain. As described by Marchese et al. (2002), usually, dendritic lakes with high shoreline development (e.g. DL > 3) will show larger numbers of microhabitats compared to lakes with low shoreline development (e.g. DL < 2)). More than of 90% of the lakes in the Paraná floodplain were originally active channels that developed into lakes as a result of abandonment of the channels (e.g., oxbow lakes) or by channel isolation caused by avulsion and by assemblages of bars and islands (Iriondo et al., 2007).

The types of hydrological connections between lotic (flowing water) and lentic (still water) environments include: (1) direct connections through mouth or erosion ditch levees or short channels, (2) indirect connections through channels of greater length, intervening lake(s) or abandoned channels that are partially covered with sediments, and (3) overbank connections, which are isolated or temporary connections that occur only during peaks of inundation phases (Iriondo et al., 2007). Thus, in alluvial floodplain lakes, large morphological and biotic spatial gradients can be observed that are related to overflow levels that determine the frequency, amplitude, duration and timing of connections to the river and thus, the exchange of nutrients (Neiff et al., 2008). Hence, it is possible to find isolated, as well as connected lakes that occur simultaneously when the river is at the level.

The temporarily connected lakes are isolated during low water season, and the shallowest habitats may dry up entirely. Limnological conditions in the lakes may be influenced by local factors such as air temperature, wind and rainfall. During rising and falling phases, the current velocities of permanent connected lakes do not exceed 0.4 m s^{-1}, and retention times vary between 7 to 10 days. In isolated lakes, the velocities and retention times change sharply when the river waters rise above the levees. Small shallow lakes have wind-driven currents, with maximum velocities of 0.02–0.03 m s^{-1} during isolation phases (Iriondo et al., 2007). The water residence time in floodplain lakes varies greatly according to the magnitude of flooding or drought phases. Thus, in extreme cases, some isolated lakes have residence time of 365 days during an isolation phase or 23 minutes during an inundation phase (Iriondo et al., 2007). Some shallow lakes may temporarily disappear by desiccation during pronounced isolation phases. The floodplain lakes usually have two types of habitats, open water and macrophyte stands. The habitat complexity provided by aquatic macrophytes (Genkai-Kato, 2007; Lucena-Moya and Duggan, 2011) promotes high species abundance and diversity (Fantin-Cruz et al., 2008).

Wetlands

Wetlands are considered among the most productive ecosystems in the world and are internationally recognized as areas with exceptional biodiversity, which makes them areas of great value for conservation (Benzaquén et al., 2017). The characterization and definition of wetlands has been widely discussed, and according to Benzaquén et al. (2017), a wetland is an environment in which temporary or permanent presence of surface or subsurface water causes its own biogeochemical flows that are different from terrestrial and aquatic environments. Distinctive features include the presence of flora and fauna adapted to these conditions, commonly hydrophytic plants and/or hydric soils or substrates with features of hydromorphism. Thus, the ecological functions of wetlands are intimately associated with hydrogeomorphic aspects (Brinson, 1993).

In the Paraná basin, most wetlands are flood-pulsing systems that oscillate between terrestrial and aquatic phases (Figure 5.2, Photo 5), and the theoretical basis for the study of these systems is based on the flood-pulse concept, or FPC (Junk et al., 1989). The flood pulse is the driving force in floodplain systems that controls both the occurrence and distribution of plants and animals, determines their traits, governs the levels of primary and secondary productivity and influences the decomposition and nutrient cycles in water and soils (Junk, 2013). The predictability of the flood pulse facilitates the adaptation of organisms to the changes between aquatic and terrestrial phases, increasing their ability to efficiently make use of periodically available resources. However, as discussed by Neiff et al. (1994), wetlands are not merely ecotones (transitional areas) between aquatic and terrestrial ecosystems, since they have their own structural and functional patterns, and the wetland biota have a structure and dynamics that is well adapted to a wide range of water availability. Large wetlands harbor specific floral and faunal species but are also periodically or permanently used by terrestrial and aquatic species from surrounding uplands and deep-water habitats. Therefore, data on species numbers collected for specific plant and animal groups in different wetlands are often incomplete and make comparative studies very difficult.

The scientific literature provides many definitions of wetlands, as well as systems for their classification, as summarized by Mitsch and Gosselink (2009). Generally, wetland classifications distinguish two groups (e.g. Neiff, 2004; Junk et al., 2014; Benzaquén et al., 2017):

- waterlogging wetlands (i.e. 'Pantanales'), which are extensive areas with sandy soils of aeolian origin or pseudo-karstic, and that are slightly concave with little slope.
- floodplains, where soil saturation is a consequence of flooding from the river, which retains excess precipitation during the rainy season and leads to periodic flooding of large interfluvial areas with periodic filling-in of depressions with water, and the lateral inundation of large areas along streams and rivers.

Large wetlands within the Paraná basin, such as the Pantanal and Iberá (Figure 5.1) are composed of many different subsystems, classified as ecosystems or habitats. The Pantanal wetland is an immense lowland of approximately 150,000 km^2 (altitude < 200 m above sea level) located in the Upper Paraguay Basin. The Pantanal landscape is a product of climatic fluctuations and environmental changes that occurred during the Late Quaternary, but geological knowledge about the formation of the Pantanal Basin and its evolution is still limited (Assine et al., 2016). The Pantanal wetland spans the international borders of Brazil, Bolivia and Paraguay, but most of this system lies within Brazil. The Pantanal is best characterized as a mosaic of seasonally inundated savanna floodplains with a spectacular diversity of flora and fauna (Pott and Pott, 2004; Junk et al., 2006).

The Iberá macrosystem occupies 12,300 km^2 within the Middle reach of the Paraná basin (Figure 5.1), primarily in Argentina, and consists of a complex of freshwater wetlands with a flooding regime of high seasonal recurrence. It originated from a section of the Paraná River during the Pliocene-Pleistocene. It is currently isolated from the flood dynamics of the river, although it likely maintains an underground connection with the river. The landscape includes temporally and permanently waterlogged areas of lentic and running waters, constituting a macro-landscape that combines forests, scrublands, grasslands, lakes, wetlands and peatlands (Neiff, 2004).

Reservoirs

Reservoirs are characterized by intermediate conditions between rivers and lakes. They can range from large-scale impoundments, such as Yacyretá and Itaipú, to small dammed rivers with a seasonal pattern of operation and water level fluctuations that are closely related to river discharge, or to constructed water bodies with pumped in-flows and out-flows. The hydrodynamics of reservoirs are greatly influenced by their operational management regime. The Brazilian part of the Paraná basin has the largest hydroelectric power potential, with more than 140 power dams that have transformed the Upper Paraná River and its tributaries into a succession of lakes (Agostinho et al., 2008). These reservoirs have eliminated floodplains, interrupting the longitudinal connectivity of rivers and damaging wetlands through the alteration of flood pulses, sediment loads and other limnological parameters (Junk et al., 2014).

Thus, construction of dams for electric power generation changes lotic systems into lentic or semi-lentic systems, promoting significant changes in flow velocity, depth, width, sedimentation, transparency and light penetration, thermal stratification and water residence times (Pagioro and Thomaz, 2002; Hoeinghaus et al., 2009; Barletta et al., 2010). Reservoirs are dynamic and heterogeneous environments in terms of spatial (vertical and longitudinal) and temporal gradients, which are characteristics related to the type of operation and the morphometry of the sites, among other factors (Nogueira et al., 2012).

In this region, an increase in water transparency has occurred due to the retention of suspended solids by the reservoirs (Henry and Maricato, 1996; Roberto et al., 2009). Noticeable local and regional limnological effects occurred due to the control and regularization of the flow (Souza Filho, 2009). These differences are the main factors responsible for the limnological and biological differences between the Upper Paraná and the lower reaches, as well as the Paraguay River, where the ecological systems are maintained by the natural river dynamics.

Ecosystems altered by man

The modifications to ecosystems by anthropogenic activities within the Paraná River Basin, ranging from deforestation for agriculture, construction of hydroelectric dams, dredging for navigation, river channeling, draining of wetlands for crops, aquaculture and urbanization. Several of these modifications promote biological invasions, species extinction, and water pollution (Daga et al., 2019). The percentage use by man in the basin is high, with less than 1% of the area designated as protected areas. The management of these protected areas is under the control of state or private organizations.

The protected areas are grouped into different categories, according to the country in which they are located and the degree of protection (Figure 5.3). These areas include strictly protected nature reserves and others that allow exploitation of natural resources under a sustainable management plan. Some protected areas are included within the RAMSAR Convention or are within Biosphere Reserves (CIC, 2017). The countries within the basin are far from achieving the commitments generated in Aichi Targets where the proportion of protected areas must reach 17% by 2020. Argentina and Paraguay have not even set targets under the Aichi Convention for protection of natural ecosystems and biodiversity. The Aichi Convention sets targets for conservation related to the UN Strategic Plan on Biodiversity (2011–2020), and signatory countries have pledged to develop management programs according to their national commitments.

Due to human activities, the biodiversity in many areas of the basin is being altered in terms of the complexity of biotic assemblages, together with reductions in the densities of populations of many species. The Paraná River Basin has the greatest density of humans and industries within the countries of Brazil, Paraguay and Argentina. Moreover, in this region, there is agriculture with heavy use of fertilizers and pesticides, destruction of riparian vegetation along tributaries and small streams, modifications to the main channel and mining activities.

La Niña–El Niño effects of the southern enso

The seasonal hydrodynamics in the basin drive the connection between the main channel and floodplain components, enabling the lateral exchange of nutrients, organic material and organisms, which drives the functioning of the ecosystem (Junk et al., 1989; Neiff, 1990; Winemiller, 2004). Depending on the river level, the lentic and lotic environments are either interconnected or isolated, both between them or within the main river channel, as illustrated in Figure 5.4 (Carvalho et al., 2001; Thomaz et al., 2007).

These effects are intensified by 'La Niña–El Niño' phenomena when the duration and the magnitude of high and low levels are so much larger. The extreme floods and droughts affect community dynamics. Extreme peaks induce modifications to nutrient dynamics, providing inflows of nutrients to flooded areas, with modifications to conductivity and levels of suspended and dissolved solids (Devercelli, 2010).

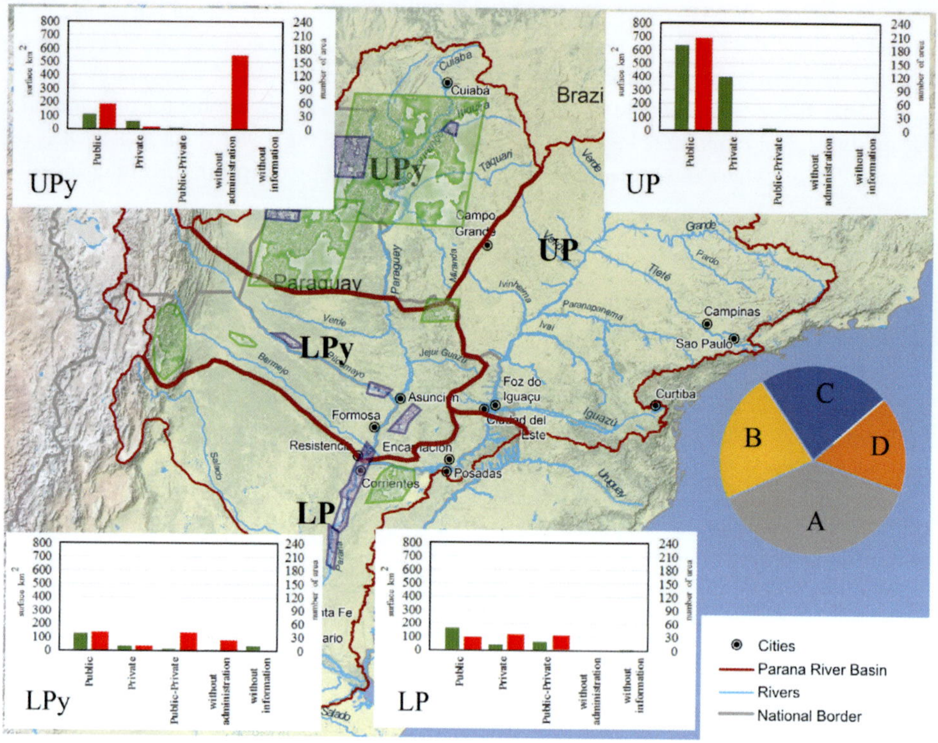

Figure 5.3 The Paraná River Basin showing the sub-basin divisions (UP: Upper Paraná, UPy: Upper Paraguay, LP: Lower Paraná, LPy: Lower Paraguay) with the location of Biosphere Reserves (green area) and Ramsar sites (blue area). Also included are the number (green bars) and surface area (red bars) of areas with some action for protection (public, private, public-private) and data on the percentage of surface area protected in relation to total surface area of each sub-basin (A: UPy, B: UP, C: LPy, D: LP). Source: Inventario de Biodiversidad de la Cuenca del Plata (PECES). 2012. Informes de consultoría de Argentina, Bolivia, Brasil, Paraguay y Uruguay.

In comparison with ordinary hydrological periods, algal density and biovolume are lower during El Niño and higher during La Niña, with the appearance of rare and sporadic phytoplankton species during the latter events (Devercelli et al., 2014). In both phases, the composition of the algal community reflects the capacity to compensate with different rates of reproduction and growth under heterogeneous conditions of water flow and light penetration. The ecological strategies of many organisms change in relation to ENSO fluctuations, including biological interactions, movements and reproduction (Winemiller, 2004; Montoya et al., 2006). Over long time-scales, the ENSO events constitute macro-factors that regulate the dispersion of freshwater fauna and the biotic interactions among species (Torres et al., 2016).

Biodiversity

The Paraná basin has developed different geographical, phytogeographic and zoogeographic regions (Morrone, 2001; Abell et al., 2008). However, there are species that are common to the

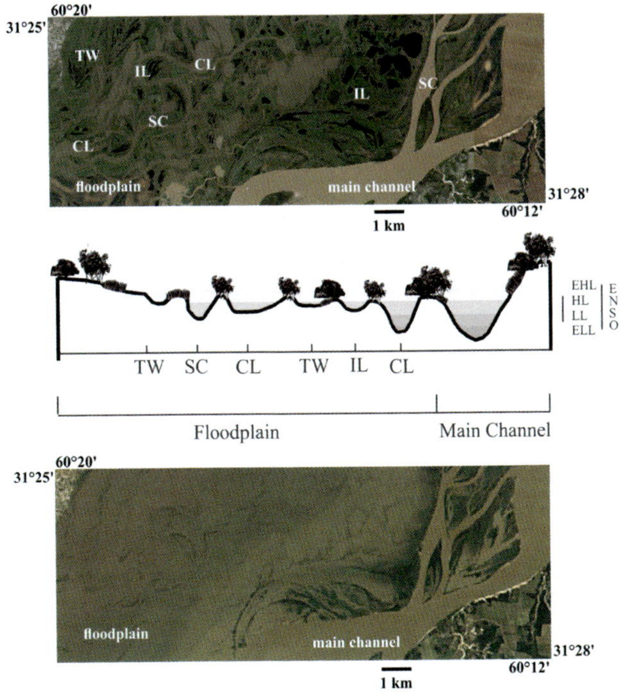

Figure 5.4 Generalized cross-section of a river-floodplain showing the low and high-water levels. CL: connected lake, EHL: extreme high level, ELL: extreme low level, HL: high level, IL: isolated lake, LL: low level, SC: secondary channel, TW: temporary wetland.

entire length of the river, establishing a continuum or biological corridor. The biogeographical zones within the basin correspond to two domains: (i) the Amazonian Dominion (including the Amazonian Provinces of Cerrado and Paranaense) in the North and (ii) the Chaco Dominion (including the Chaqueña, Espinal and Pampeana Provinces) in the South (Cabrera and Willink, 1973). There is considerable continuity of vegetation in its physiognomy, even though the rivers of the basin cross ecozones of tropical rain forest, savanna, steppes or spinal.

The terrestrial flora in the Upper Paraná River is comprised of 97 families with 450 species of phanerogams, located mainly in riparian areas. Moreover, 86 species of aquatic plants have been described in Brazil territory (Thomaz et al., 2004) and 114 in the Argentine stretch of this river (Neiff et al., 2014). In the Pantanal region of the Upper Paraguay basin, Pott and Pott (2000) described 247 species of aquatic plants (Figure 5.5). This wetland is a savanna that is seasonally flooded by the Paraguay River and includes several marginal fluvial fans/megafans. In the same region, there are various vegetation types, such as deciduous forest and Cerrado woodland in transition with grassland or riparian forest at the edges of shallow lakes.

In the Middle and Lower stretches of the Paraná, the aquatic macrophytes include around 100–114 species, with the emergent species being the more abundant group (Figure 5.5). Other aquatic macrophytes include free-floating, rooted submerged, floating-leaved and free submerged groups (Neiff et al., 2014; Schneider et al., 2018, 2019). The terrestrial vegetation is divided into seasonal deciduous forest and seasonal semi-deciduous forest, including islets and riperian forest (Figure 5.6), pioneer woodlands (including *Vochysia divergens*), scrub (thorn scrub,

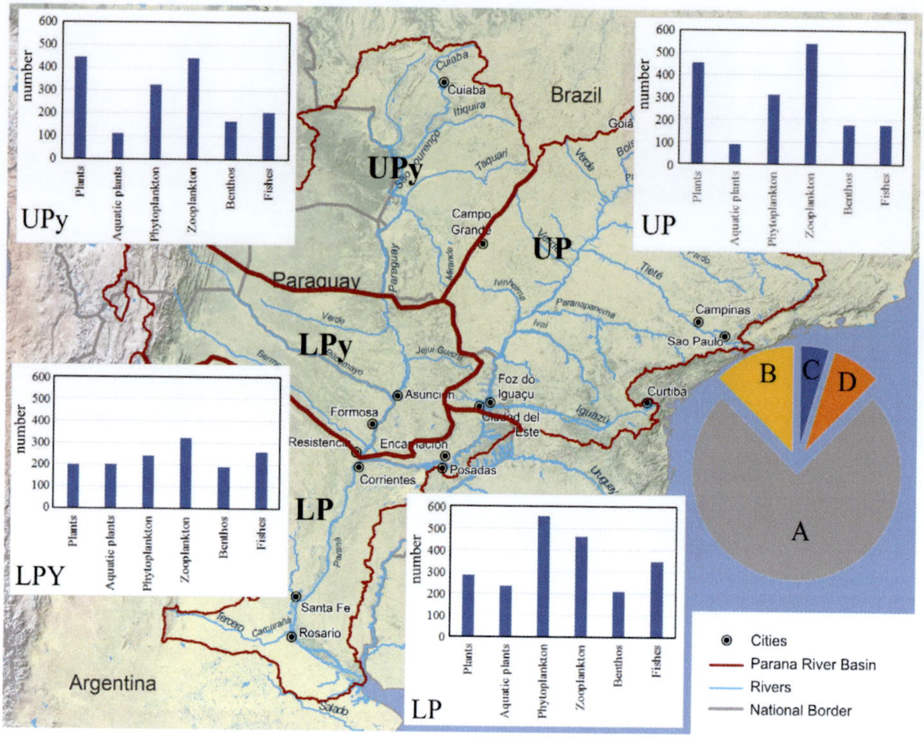

Figure 5.5 Paraná River Basin showing sub-basin divisions (UP: Upper Paraná, UPy: Upper Paraguay, LP: Lower Paraná, LPy: Lower Paraguay), the number of species in each sub-basin for terrestrial plants, aquatic plants, phytoplankton, zooplankton, benthos and fishes (bars) and the percentage in the basin of other vertebrates (A: birds, B: mammals, C: amphibians, D: reptiles).

shrub thickets, including *Couepia uiti* and *Licania parviflora*), riparian grass (giant sedge, fire flag), swamp, floating meadow, savanna or 'Cerrado' (including Cerrado woodland, floodable mono-dominant savannas and grasslands) and steppic savanna or 'Chaco'.

The aquatic vegetation is configured in a zonation pattern according to water depth. The majority of the vegetation is emergent and amphibious plants. Along the rivers, there are rooted floating mats of anchored water hyacinth *Eichhornia azurea*, while stands of emergent plants develop in the shallow margins (*Polygonum* spp.) as shown in Figure 5.6. The submerged plant genera (*Cabomba*, *Ceratophyllum*, *Najas*) occur mostly in shallow lakes and streams away from turbid waters. Free-floating plant species include duckweed (Lemnoideae), floating ferns (*Salvinia* spp.), mosquito ferns (*Azolla* spp.), water lettuce (*Pistia stratiotes*) and water hyacinth (*Eichhornia crassipes*), as illustrated in Figure 5.6. Rooted and floating hydrophytes include Victoria lily (*Victoria* spp.), which is the largest aquatic macrophyte that grows in oxbow lakes of the Paraguay and Paraná rivers, and water lilies (*Nymphaea* spp.) that occur in seasonal and permanent lakes. These species show changes in cover, depending on the topographic position. The free-floating plants occupy intermediate positions or show displacements towards the bank or limnetic water. Emergent plants occur around wetlands, such as the giant sedge (*Cyperus giganteus*), pickerel weed (*Pontederia parviflora*) and fire flag (*Thalia geniculata*). Amphibious plants are found in floodable grasslands and savannas, with many grasses, sedges and bush morning glory (*Ipomoea carnea*), which grow either in flooded or dried ground (Pott and da Silva, 2016). The macrophyte biomass fluctuates between 4.18 and 59.04 tonnes wet weight ha^{-1}, and between 0.30 and 8.67 tonnes dry weight ha^{-1}, produced mainly by *P. repens*, *P. stratiotes* and *L. peploides* (60 to 80%), and, to lesser extent (i.e. 10% or lower) by *A. caroliniana*, *E. anagalis*, *Limnobium laevigatum*, *P. elephantipes*, *H. amplexicaulis*, *E. azurea*, *E. crassipes*, *P. rotundifolia*, *S. herzogii*, *Scirpus cubensis* and *Utricularia* spp. (Iriondo et al., 2007).

The numbers of species within various taxonomic groups are summarized in Figure 5.5, according to the numbers of species in the sub-basins of the Paraná River system. The phytoplankton of the entire Paraná basin includes about 800 taxa. The largest contribution to species richness corresponds to Chlorococcales, and this group is followed in importance by Bacillariophyceae in the main channel and Euglenophyceae in the floodplains. The dynamics of phytoplankton populations is affected by hydro-sedimentological regime that allows lateral exchange between floodplain lakes with higher algal accumulation and their lotic environments. The density is higher at the end of the low water period in the main channel and in the floodplain lakes (Thomaz et al., 2004), as the higher water levels produce dilution effects with lower densities (Thomaz et al., 2007).

In the Upper Paraná River, 300 taxa of phytoplankton have been recorded in floodplain lakes and 177 taxa in the main channel. Bacillariophyceae, Chlorophyceae and Cyanophyceae are the most important taxonomic groups (Rodrigues et al., 2009). The Cyanophyceae are frequently found in floodplain lakes, producing blooms of some species like *Anabaena* spp. and *Microcytis aeruginosa* during the low water phase. However, the greatest number of algae are found in Upper Paraguay in the Pantanal, with almost 600 species (Zalocar de Domitrovic, 2005). Unlike many other rivers, algal species richness declines towards the mouth of the

Figure 5.6 A shallow lake in the floodplain of the Middle stretch of the Paraná River, showing the diversity of aquatic and terrestrial environments based on vegetation. The photograph was provided from the Archive of the Instituto Nacional de Limnología, Argentina.

Paraná, where there are only about 200 species (Zalocar de Domitrovic, 2005). In the tributaries with headwaters in the Andes mountain, the richness and densities of phytoplankton are low (e.g. 210 species and 176 individuals ml^{-1}), characterized by *Chlamydomonas, Dunaliella, Klebsormidium, Spirogyra, Zygogonium, Ulothrix*, among others. The abundances of Bacillariophyta and Cyanophyta increase with snow melt in the Andes each spring (Moraña, 2013).

In the Lower Paraná River, there are 308 species of phytoplankton, characterized by a dominance of centric diatoms together with other taxonomic groups, such as coccal green algae, and many sporadic species (Devercelli et al., 2016). The main channel is characterized by low algal densities, chlorophyll concentration and specific richness. Bacillariophyceae (mostly represented by *Aulacoseira granulata* and its bioforms) dominate (i.e. 50%), followed by Chlorophyceae and Cyanobacteria, which originate from the Upper Paraná River. In a more southern stretch, Chlorophyceae and Cryptophyceae share alternatively sub-dominance. In floodplain lakes phytoplankton are dominated by Cryptophyceae, Euglenophyceae, Chlorophyceae and Chrysophyceae, while diatom species are mainly tychoplanktic or potamoplanktic during the flood phase (Iriondo et al., 2007).

Some zooplankton species are common to all the large South American basins, including the Orinoco, Amazon and Rio del Plata, of which the Paraná River is part, which is recognized as being part of the same biogeographical Guyano-Brazilian sub-region. In lotic environments, rotifers and copepod nauplii dominate with low densities, while lentic environments show higher zooplankton densities and richness with greater proportions of adult copepods (Iriondo et al., 2007). Moreover, littoral areas have higher species richness, which is associated with greater habitat heterogeneity and resource availability (Lansac-Tôha et al., 1997). The water level affects similarly the composition of several zooplankton groups, showing higher richness during the high-water phase. However, more abundance is associated with periods of low water velocity, longer retention times and greater food availability.

In the Upper Paraná, zooplankton is represented by 541 taxa, with rotifers being most abundant, followed by testate amoeba species, cladocerans and copepods (Lansac-Tôha et al., 2009). The richness varies with habitat, where lakes support the highest rotifer numbers, but testate amoeba occur with greater diversity in rivers, while the diversity of microcrustaceans is similar across different habits (Garcia et al., 1998). The zooplankton density in the Paraguay River decreases from north to south (Frutos et al., 2006). When comparing zooplankton abundance in Upper Paraná with the Lower Paraná, the latter shows higher organism densities, due to inputs from the Paraguay River and greater development in the middle stretch floodplain (Sendacz, 2001).

Zooplankton species richness does not vary greatly over the middle stretch of the river over a distance of more than 600 km (Iriondo et al., 2007). In this stretch, there are 230 rotifer species, 125 cladocerans and 87 of copepods in the main channel and floodplain environments (Paggi, 2004). The highest diversity of rotifers is in the genera *Brachionus, Lecane* and *Trichocerca* with abundant cosmopolitan species, but with endemic species to the Neotropical region (e.g. *Brachionus dolabratus, B. mirus, B. zahniseri, Lecane proiecta, Hexarthra intermedia braziliensis, Testudinella ohlei* and *Lepadella donneri*). Cladocerans have the highest diversity in the genera *Diaphanosoma*, followed by *Ceriodaphnia, Moina, Bosmina* and *Daphnia*. Most of the cladoceran species are endemic to the Neotropical region, including the Caribbean and southeastern parts of North America (e.g *Diaphanosoma brevirreme, D. fluviatile, Moina minuta, M. reticulata, Daphnia gessneri, Bosmina hagmani* and *B. huaronensis*). The calanoid copepods are characterized by *Notodiaptomus*, which are endemic of a Neotropical region, while among cyclopoid copepods, *Ectocyclops, Eucyclops, Mesocyclops, Microcyclops* and *Thermocyclops* are the most diverse genera (Iriondo et al., 2007, Perbiche-Neves et al., 2014). One of the most ecologically important events in the Paraná River Basin was the invasion of the Asiatic bivalve *Limnoperna fortunei*, which is associated with declines in zooplankton abundance (mainly *Keratella* spp.) and reduced levels of chlorophyll a in secondary floodplain channels (Rojas Molina and José de Paggi, 2008).

The Upper and Lower Paraná River have similar benthic species assemblages (Marchese et al., 2002; Thomaz et al., 2004), and similar benthos occurs in the Upper and Lower Paraguay River (Marchese et al., 2005). The benthic composition is stable along a stretch of approximately 3,500 km and 18° of latitude, with sand-adapted assemblages of small-sized benthic organisms. These are like the bottom fauna that inhabit mobile sand-bottom rivers worldwide, such as the White Nile in Africa, the Athabasca River in Canada, the Amur River and Ganges River (Wantzen et al., 2014). Studies on Paraguay-Paraná River and floodplain support the concept that benthic invertebrate diversity increases laterally, from the main channel to floodplain habitats and is controlled by physical factors, such as turbulence, flow velocity, hydraulic parameters and turbidity (Iriondo et al., 2007, Wantzen et al., 2014). The benthic diversity and species evenness are higher in lakes connected to the main channel or to secondary channels (Zilli and Marchese, 2011). The similar number of taxa (i.e. 150) over the entire Paraná basin demonstrates that the flood pulse produces and maintains a diverse and dynamic structure with great biodiversity, despite the stress of these transitional periods. To compensate for the drastic changes in densities and species richness between aquatic and terrestrial phases, the r-strategist organisms have rapid growth, early maturity and high rates of reproduction. However, K-strategists are favored in other periods because of their competitive capacity. Spatial heterogeneity, habitat structure, temporal instability and high productivity determine the greater importance of lateral dimensions than longitudinal dimensions in determining benthic communities in large rivers.

In the Upper Paraná, benthic invertebrates are very abundant, with more of 100 taxa of Chironomidae alone, and with Oligochaeta, Bivalvia and Crustacea also being important taxa. Chironomidae dominate in the river margins, while the main channel is inhabited by interstitial organisms, especially nematodes, Harpacticoidea and Oligochaeta. The difference between habitats is attributed to the sediment texture, water velocity and particulate organic matter content (Stevaux and Takeda, 2002; Petsch et al., 2015). In the main channel of the Paraná River, *Narapa bonettoi* (Oligochaeta) is dominant and often the only benthic species, associated with lower densities of *Myoretronectes paranaensis*, *Itaspiella parana* (Turbellaria), *Haplotaxis aedeochaeta* (Oligochaeta), *Tobrilus* sp. (Nematoda), *Potamocaris* spp. (Harpacticoid crustacea), *Parachironomus* sp. and *Lopescladius* sp. (Chironomidae). A remarkable longitudinal change in benthos was observed in the basaltic rock beds in the Apipé rapids of the Upper Paraná, with the appearance of Porifera species, such as *Oncosclera navicella*, *O. schubarti*, *Trochospongilla repens*, *T. ringueleti* and *Corvospongilla seckti* (Iriondo et al., 2007).

Within river-floodplain habitats, there are more than 150 taxa of benthic invertebrates, without including Turbellaria, Nematoda and Crustacea, with a higher proportion of oligochaetes in the channels and lakes, and greater numbers of insects within the temporary wetlands. In temporary wetlands, in addition to many oligochaete species, there is a higher proportion of Insecta (mainly Coleoptera, *Berosus* spp., Chironomidae, *Polypedilum* spp., *Ceratopogonidae* spp., *Dolychopodidae* spp., *Tipulidae* spp., Odonata *Aphylla* sp.), as well as Ostracoda (Iriondo et al., 2007; Zilli and Montalto, 2011).

The Paraná River, as with other large South American rivers (e.g., Amazon, Orinoco, Magdalena, etc.), is characterized by a high species richness and production of macrophytic plants. These macrophytes generate a great flux of organic matter to the bottom, which produces hypoxic and/or anoxic conditions, mainly during the low water period and the beginning of the flooding phase. The biotic assemblages have adjusted to these extreme conditions for survival. Among the aquatic plant community are abundant periphytic communities, composed of algae, micro- and macro-crustaceans, mollusks and insect larvae, in addition to vertebrates (fishes, amphibious, reptiles, birds and mammals) that interact for space, refuge and food. The Upper Paraná, Lower Paraná and Paraguay rivers have insects and crustaceans as dominant organisms that are associated with the roots of macrophytes (Souza-Franco and Takeda, 2000; Poi de Neiff and Neiff, 2006).The invertebrate fauna that occur in association with *Eichhornia* sp. macrophytes distributed from the Pantanal to the confluence of Paraguay River with Paraná River are similar in densities and richness (Poi de Neiff, 2003). The variation of assemblage of populations that live in roots is strongly influenced by the horizontal flow of water. Moreover, the functional groups of invertebrates associated with these root systems include herbivores and shredder invertebrates (Poi de Neiff, 2003).

With regards to fishes, the biotic complexity of fluvial system has several axes, which can be observed when analyzing fish movements, which can be in the direction of the river flow for reproductive migrations, and transverse to the flow in search of refuge and food (Welcomme and Halls, 2002). Some fish travel more than 600 km in their reproductive migrations (Bonetto, 1986). In the stretch of Upper Parana, the ichthyofauna is composed of 170 species. The upper reaches of the river contain important spawning areas for migratory fishes of the basin. 'Pacu' (*Piaractus mesopotamicus*), 'armado' (*Rhinoderas d'orbinyi*) and 'pintado' (*Pseudoplatystoma corruscans*) are the most readily recognizable fish species.

On the floodplains, which offer a high diversity of refuge areas and abundant food resources, the number of fish species is relatively high, with 103 species recorded in lakes and 101 species in the channels that connect them to the rivers. The fishes of the floodplain lakes are composed of small species and the young of larger species that utilize these habitats during their early development. Characiforms dominate numerically (55%) over siluriforms (22%) in the lakes. The species number is also much higher in the upper stretch relative to the lower stretch of the Paraguay-Paraná system. In the Pantanal, there are more than 350 species, while 230 species have been recorded for the Paraná at the mouth of the Río de la Plata (López et al., 2002). However, if only the species captured in the main channel are considered, there are more species in Lower Paraguay than in the Upper Paraguay (Canon Verón, 2005). This implies that a large part of the fishes that inhabit the Pantanal wetland do not use the river as a migratory connection or are isolated from the main flow.

A wide diversity of other vertebrates, including amphibians, reptiles, birds and mammals are included in the inventory of species in the Paraná Basin. Twenty-two species of amphibians have been recorded, with Hylidae and Leptodactylidae being the most abundant groups. Moreover, 37 reptile species have been recorded, with the Colubridae Family (n=16 species) being the richest in species diversity. The reptiles include three crocodile species, a small number of poisonous snakes and many non-poisonous species. Furthermore, 58 bird families have been recorded, with the Passerine order being the most abundant group, with 350 species. This region is one of the richest due to the influence of several biogeographic events that occurred in the past. At least 60 mammal species have been documented, with the Phyllostomatidae family being the richest in numbers of species, followed by Felidae, Cervidae and Cricetidae. Among the notable mammals in the area are the great predator cats, the puma (*Felis concolor)* and jaguar (*Panthera onca*). Moreover, there are large populations of capybara (*Hydrochaeris hidrochaeris*), wild pig (*Tayassu tajacu*), wild dog (*Cerdocyon thous*), maned wolf (*Canis brachyurus*), tapir

(*Tapirus terrestris*), in addition to large populations of mash deer, anteaters, howler monkeys and other monkey species and coati.

Conservation

More than 10% of all described species on earth are concentrated in freshwater ecosystems, although only 1% of the Earth's surface consists of aquatic environments, making these systems a biodiversity hotspot. What's more, beneficial services provided by freshwater ecosystems have been estimated to be worth more than $29 trillion in US dollars a year (Kawai and Cumberlidge, 2016). Despite the immense value of freshwater ecosystems as biodiversity hotspots and as providers of ecosystem services, these systems are increasingly impacted by threats such as water pollution, flow modification, habitat destruction or degradation and invasion by exotic species. The rates of loss of habitats and species in freshwater ecosystems are larger than those seen in either terrestrial or marine ecosystems (Dudgeon et al., 2006)

Although losses of habitat and species have not been so intense in the Paraná River Basin as a whole, some areas or zones have been substantially impacted. Unfortunately, the effort to determine the list of species at risk in the Paraná basin is limited to certain groups and has been fragmented, due to the diversity of species and environments and the lack of political will in each country. Pressures on freshwater resources in the Paraná River and the need to monitor the basin is growing daily, with the increases in population and the demand for safe drinking water, improved sanitation, food, energy and industrial products. However, the questionable environmental accounting methods in use for environmental impact assessments of development projects fail to factor in the losses of valuable ecosystem services provided by freshwater systems and often underestimate the true costs of the detrimental impacts of development. The increasing exploitation of freshwaters for human use fails to recognize the needs of the species that live there, and equally does not value the services that these species collectively provide to ensure the sustainable functioning of the ecosystems.

The Paraná River Basin must be observed in its integrity, recognizing both the main channel and the floodplains. The characteristics of the basin and its tributaries are integral to understanding the types of land use. This in turn has interactions with the hyporheic zone, such that alluvial aquifers associated with wetlands are key factors in the recharge of groundwaters.

There is a need within the Paraná River Basin to establish a network of carefully managed protected areas that are large enough to preserve natural forests in drainage basins that will maintain water quality in the rivers, streams, lakes and wetlands. This in turn would require the establishment of more nature reserves and national parks, and the development of integrated species conservation action plans, especially for species that are threatened with extinction. Moreover, the settlement routes of South American fauna have shown the fundamental role of rivers as biological corridors (Ringuelet, 1975, López et al., 2002, Menni, 2004) that have contributed to the wide dispersion of most species.

The International Union for Conservation of Nature (IUCN) Red List assessments of biodiversity includes more than 1200 species that have been listed as endangered, including plants, invertebrates and vertebrates (Figure 5.7). The threatened species are mainly those with a restricted range (stenotopic), endemic to certain regions, or of economic interest to man, and the habitat of these species has been impacted by deforestation, alterations to drainage patterns, pollution and hunting or fishing without effective controls. Among them, habitat specialists have small populations and are restricted in their distributional ranges and are especially susceptible to localized anthropogenic habitat disturbances. It is also very important to know the

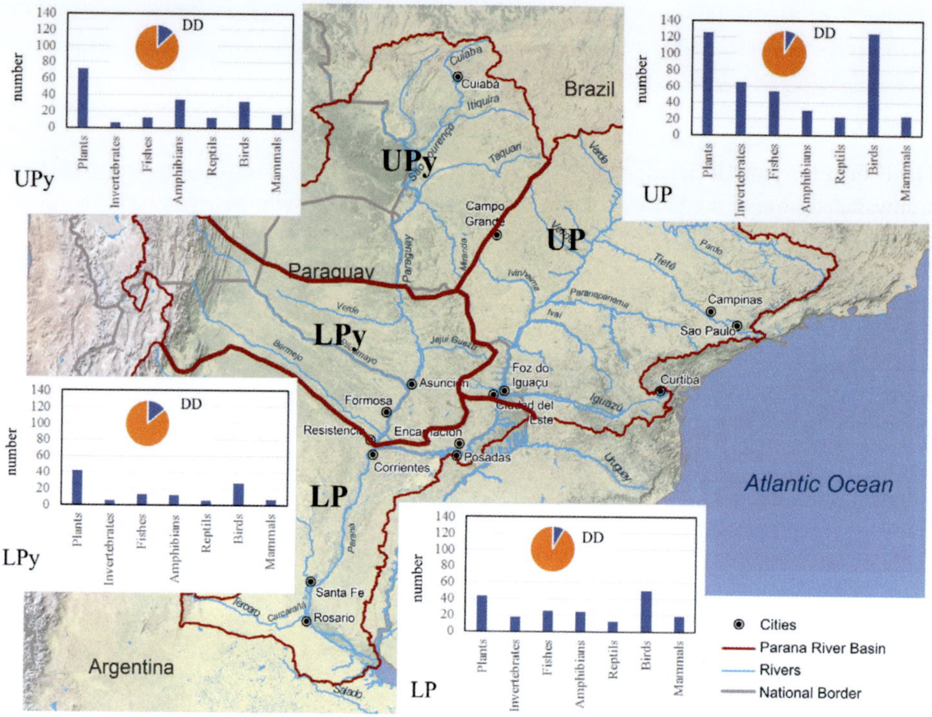

Figure 5.7 Paraná River Basin showing the sub-basin divisions (UP: Upper Paraná, UPy: Upper Paraguay, LP: Lower Paraná, LPy: Lower Paraguay) and the number of endangered species of plants, invertebrates and vertebrates (bars), and the data deficient percentage (DD). Color coding: blue: plants, orange: animals. Source: UICN (www.iucnredlist.org).

effects of introduced species and their capacity to threaten native biodiversity. In addition, species that host parasites may affect various domestic and wildlife species, and even humans.

Many species at risk within the Paraná basin are not known beyond their original description, and their actual geographic range (other than the type of habitat) is unclear due to inadequate on-the-ground knowledge (Figure 5.7). Once a species has been initially assessed as Data Deficient (DD) it is not included in any further analyses, because data necessary for inclusion on the IUCN Red List are lacking. High numbers of DD species are therefore a cause for concern because this designation may be masking the true number of threatened species. IUCN Red List threat levels could increase significantly when new field data enable their inclusion in future assessments.

Multiple gaps in the exploration of freshwater diversities in the Paraná basin make it likely that more intensive surveying will not only add to the number of known species, it will also add to the number of threatened and DD species. The eventual inclusion in the Red List of little-known species currently treated as DD is likely to alter the current levels of biodiversity threat. Another factor that adds to the number of known species (and to the number of threatened species) is a lack of recent samples for the discovery of cryptic species using DNA sequence data. The new capacity offered by state-of-the-art instrumentation for molecular analysis, as well as new morphological approaches will allow detailed evaluations that can reveal cryptic species. Taken together with present high numbers of DD species, it seems highly likely that the true number of threatened species in the Paraná basin may be much higher than originally estimated.

Conclusions

The Paraná River is characterized by high diversity of aquatic environments that span from the headwaters to the mouth at the discharge into the Rio de La Plata estuary. The main categories of these environments include the main channel, secondary channels, connected and isolated shallow lakes and wetlands. The main tributaries add one more aspect to the diversity of biotic and abiotic conditions. These include the Paraguay, Bermejo, Pilcomayo and Salado rivers. This produce a complex mosaic of biodiversity within the Paraná basin.

To the variability in aquatic environments, there is the added complexity of the different biogeographic regions, as the Paraná River and its tributaries pass through tropical and subtropical forests, forests in galleries, large wetlands, foothill forested strips (i.e. yungas), grasslands, arid lowlands (i.e. chaco), mountains; all that occur in a longitudinal axis over almost the full extent of the South American continent. The headwaters of the Paraná River and Paraguay River are located in the tropical north and the fluvial system then moves south through subtropical and then temperate climates, contributing to high biodiversity.

The latitudinal and longitudinal development of the basin promotes the spatial and temporal heterogeneity that characterize the Paraná River. The dynamic of connections among the different aquatic environments contribute to the high biodiversity. This diversity is supported by the connectivity among the environments, mainly between the main channel and floodplains.

The biological diversity includes from bacteria to mammals, but the aquatic vegetation represents a main group that permit structuring of communities where they find refuge, food and the energy y matter is autochthone into the system.

The Upper Paraná is characterized by a high level of impacts from humans (e.g. high urban density, hydroelectric development), but many areas on the Paraguay River and in the Middle Reach and Lower Reach of the Paraná River are relatively untouched by human development. The scarce human intervention in some parts of the Paraná basin ensures that biological diversity that developed evolutionarily has been preserved. However, the unequal participation of the governments in studies of biodiversity within the basin makes it difficult to categorize the species according to the degree of vulnerability to anthropogenic impacts. The loss of heterogeneity of the system is the main threat to biodiversity and the maintenance of healthy ecosystems within the basin.

Acknowledgements

We are grateful to the research staff, students and technician of Instituto Nacional de Limnología (INALI-CONICET-UNL) for their collaboration. We would like to thank the Consejo Nacional de Investigaciones Científicas y Técnicas (CONICET) and the Universidad Nacional del Litoral (UNL) for the financial support.

References

Abell, R., Thieme, M.L., Revenga, C., Bryer, M., Kottelat, M., Bogutskaya, N., ... Petry, P. 2008. Freshwater ecoregions of the world: A new map of biogeographic units for freshwater biodiversity conservation. *Bioscience* 58:403–414. https://doi.org/10.1641/B580507

Agostinho, A.A.,, Pelicice, F.M., Gomes, L.C. 2008. Dams and the fish fauna of the Neotropical region: Impacts and management related to diversity and fisheries. *Brazilian Journal of Biology* 68:1119–1132.

Amsler, M., Gaudin, H. 1994. La superposición de dunas y el transporte de la carga de fondo en el Río Paraná. In: *Anales del XV Congreso Nacional del Agua (La Plata, Argentina)*. Universidad Nacional de La Plata, La Plata (Buenos Aires), Argentina, pp. 59–65.

Amsler, M.L., Drago, E.C. 2009. A review of the suspended sediment budget at the confluence of the Paraná and Paraguay Rivers. *Hydrological Processes* 23:3230–3235.

Assine, M.L., Merino, E.R., Pupim, F.N., Warren, L.V., Guerreiro, R.L., McGlue, M.M. 2016. Geology and Geomorphology of the Pantanal Basin. In: Bergier, I., Assine, M.L. (eds.), *Dynamics of the Pantanal Wetland in South America*, Springer, New York, pp. 23–50.

Barletta, M., Jaureguizar, A.J., Baigun, C., Fontoura, N.F., Agostinho, A.A., Almeida-Val, F.V.M., Val, A.L., Torres, R.A., Jimenes-Segura, L.F., Giarrizzo, T., Fabre, N.N., Batista, V.S., Lasso, C., Taphorn, D.C., Costa, M.F., Chaves, P.T., Vieira, J.P., Correa, M.F.M. 2010. Fish and aquatic habitat conservation in South America: A continental overview with emphasis on neotropical systems. *Journal of Fish Biology* 76:2118–2176.

Benzaquén, L., Blanco, D., Bo, R., Kandus, P., Lingua, G., Minotti, P., Quintana, R. (eds.) 2017. *Regiones de Humedales de la Argentina*. Ministerio de Ambiente y Desarrollo Sustentable, Fundación Humedales/ Wetlands International, Universidad Nacional de San Martín and Universidad de Buenos Aires, Argentina.

Blettler, M., Amsler, M., Ezcurra de Drago, I., Espinola, L., Eberle, E., Paira, A., Best, J., Parsons, D., Drago, E. 2014. The impact of significant input of fine sediment on benthic fauna at tributary junctions: A case study of the Bermejo-Paraguay River confluence, Argentina. *Ecohydrology* 8:340–352.

Blettler, M.C.M., Amsler, M.L., Espinola, L.A., Rabuffetti, A.P., Ezcurra de Drago, I., Drago, E.C. 2019. Ecological relevance of anabranch confluences: Hydrological and morphological effects. *Ecohydrology* 12:e2068.

Bonetto, A.A. 1986. The Paraná River System. In: Walker, K.F., Davies, B.R. (eds.), *The Ecology of River Systems*, Junk Publishers, Dordrecht, The Netherlands, pp. 541–551.

Bonetto, A.A., Wais, I.R. 1995. Southern South American streams and rivers. In: Cushing, C.E., Cummins, K.W., Minshall, G.W. (eds.), *River and Streams Ecosystems*. Elsevier, Amsterdam, pp. 257–293.

Brinson, M.M. 1993. *A Hydrogeomorphic Classification for Wetlands. Wetlands Research Program Technical Report WRP-DE-4.*

Cabrera, A., Willink, A. 1973. *Biogeografía de América Latina. Programa Regional de Desarrollo Científico y Tecnológico.* Departamento de Asuntos Científicos. Secretary General, Organization of American States (OAS), Washington, D.C., USA.

Camilloni, I., Barros, V. 2003. Extreme discharge events in Paraná River and their climate forcing. *Journal of Hydrology* 278:94–106.

Canon Verón, M.B. 2005. *Patrones de distribución y abundancia de peces en el sistema de confluencia de los ríos Paraná y Paraguay*. Doctoral thesis, Universidad Nacional de Córdoba, Argentina.

Carvalho, P., Bini, L.M., Thomaz, S.M., Oliveira, L.Q., Robertson, B., Tavechio, W.L.G., Darwisch, A.J. 2001. Comparative limnology of South American floodplain lakes and lagoons. *Acta Scientiarum* 23:265–273.

Comité Intergubernamental Coordinador de los Países de la Cuenca del Plata- CIC. 2017. *Ecosistemas acuáticos en la Cuenca del Plata*. Estados Unidos, Organization of American States (OAS), Ciudad Autónoma de Buenos Aires.

Daga, V.S., Azevedo-Santos, V.M., Pelicice, F.M., Fearnside, P.M., Perbiche-Neves, G., Paschoal, L.R.P., Cavallari, D.C., Erickson, J., Ruocco, A.M.C., Oliveira, I., Padial, A.A., Vitule, J.R.S. 2019. Water diversion in Brazil threatens biodiversity: Potential problems and alternatives. *Ambio* 67:23–36.

Devercelli, M. 2010. Changes in phytoplankton morpho-functional groups induced by extreme hydroclimatic events in the Middle Paraná River (Argentina). *Hydrobiologia* 639:5–19.

Devercelli, M., Scarabotti, P., Mayora, G., Schneider, B. Giri, F. 2016. Unravelling the role of determinism and stochasticity in structuring the phytoplanktonic metacommunity of the Paraná River floodplain. *Hydrobiologia* 764:139–156.

Devercelli, M., Zalocar de Domitrovic, Y., Forastier, M., Zaburlin, N. 2014. Phytoplankton of the Paraná River Basin. In: Tell, G., Izaguirre, I., O'Farrell, I. (eds.), Freshwater phytoplankton from Argentina,. *Fundamental and Applied Limnology* 65(Special Issue):39–65.

Drago, E., Ezcurrra de Drago, I., Oliveros, O., Paira, A. 2003. Aquatic habitats, fishes and benthic assemblages of the Middle Paraná River. *Amazoniana* 17:291–341.

Drago, E.C. 1990. Geomorphology of large alluvial rivers: Lower Paraguay and Middle Paraná. *Interciencia* 15:378–387.

Drago, E.C., Amsler, M.L. 1998. Bed sediment characteristics in the Paraná and Paraguay rivers. *Water International* 23:174–183.

Drago, E.C., Paira, A.R., Wantzen, K.M. 2008. Channel-floodplain geomorphology and connectivity of the Lower Paraguay hydrosystem. *Ecohydrology & Hydrobiology* 8:31–48.

Dudgeon, D., Arthington, A.H., Gessner, M.O., Kawabata, Z.I., Knowler, D.J., Leveque, C., Naiman, R.J., Prieur-Richard, A.H., Soto, D., Stiassny, M.L.J., Sullivan, C.A. 2006. Freshwater biodiversity: Importance, threats, status and conservation challenges. *Biological Reviews* 81:163–182.

Fantin-Cruz, I., Tondato, K.K., Penha, J.M.F., Mateus, L.A.F., Girard, P., Fantin-Cruz, R. 2008. Influence of fish abundance and macrophyte cover on microcrustacean density in temporary lagoons of the Northern Pantanal, Brazil. *Acta Limnologica Brasiliensia* 20:339–344.

Frutos, S.M., Poi de Neiff, A.S.G., Neiff, J.J. 2006. Zooplankton of the River: A comparison between sections and hydrological phases. *Annales de Limnologie* 42:277–288.

Garcia, A.P.P., Lansac-Tôha, F.A., Bonecker, C.C. 1998. Species composition and abundance of rotifers in different environments of the floodplain of the Upper Paraná River, Brazil. *Revista Brasileira de Zoologia* 15:327–343.

Genkai-Kato, M. 2007. Macrophyte refuges, prey behaviour and trophic interactions: Consequences for lake water clarity. *Ecology Letters* 10:105–114.

Henry, R., Maricato, F.E. 1996. Sedimentation rates of tripton in Jurumirim Reservoir (São Paulo, Brazil). *Limnologica* 26:15–25.

Hoeinghaus, D.J., Agostinho, A.A., Gomes, L.C., Pelicice, F.M., Okada, E.K., Latini, J.D., Kashiwaqui, E.A.L., Winemiller, K.O. 2009. Effects of river impoundment on ecosystem services of large tropical rivers: Embodied energy and market value of artisanal fisheries. *Conservation Biology* 23:1222–1231.

Iriondo, M., Paggi, J.C., Parma, J. (eds.) 2007. *The Middle Paraná River: Limnology of a Subtropical Wetland.* Springer, New York.

Junk, W.J. 2013. Current state of knowledge regarding South America wetlands and their future under global climate change. *Aquatic Science* 75:113–131.

Junk, W.J., Bayley, P.B., Sparks, R.E. 1989. The flood pulse concept in river-floodplains systems. *Canadian Special Publication Fisheries Aquatic Science* 106:110–127.

Junk, W.J., Nunes Da Cunha, K., Wantzen, K.M., Petermann, P., Strussmann, C., Marques, M.I., Adis, J. 2006. Biodiversity and its conservation in the Pantanal of Mato Grosso, Brazil. *Aquatic Science* 68:278–309.

Junk, W.J., Piedade, M.T.F., Lourival, R., Wittmann, F., Kandus, P., Lacerda, L.D., Bozelli, R.L., Esteves, F.A., Nunes da Cunha, C., Maltchik, L., Schöngart, J., Schaeffer-Novelli, Y., Agostinho, A.A. 2014. Brazilian wetlands: their definition, delineation, and classification for research, sustainable management, and protection. *Aquatic Conservation Marine and Freshwater Ecosystems* 24:5–22 https://doi.org/10.1002/aqc.2386

Kawai, T., Cumberlidge, N. (eds.) 2016. *A Global Overview of the Conservation of Freshwater Decapod Crustaceans.* Springer Verlag, Heidelberg, Germany.

Lansac-Tôha, F.A., Bonecker, C.C., Velho, L.F.M., Lima, A.F. 1997. Comunidade zooplanctônica. In: Vazzoler, A.E.A.M., Agostinho, A.A., Hahn, N.S. (eds.), *A planície de inundação do alto rio Paraná: Aspectos físicos, químicos, biológicos e sócio-econômicos.* EDUEM, Maringá, Argentina, pp. 117–155.

Lansac-Tôha, F.A., Bonecker, C.C., Velho, L.F.M., Simões, N.R., Dias, J.D., Alves, G.M., Takahashi, E.M. 2009. Biodiversity of zooplankton communities in the Upper Paraná River floodplain: Interannual variation from long-term studies. *Brazilian Journal of Biology* 69:539–549.

Lima, J.E.F.W., Lopes, W.T.A., Carvalho, N.O., Vieira, M.R., Silva, E.M. 2005. Suspended sediment fluxes in the large river basins of Brazil. In: Walling, D.E., Horowitz, A.J. (eds.), *Sediment Budgets 1. Foz do Iguaçu.* International Association of Hydrological Sciences (IAHS), Wallingford, UK, pp. 355–363.

López, H.L., Morgan, C.C., Montenegro, M.J. 2002. *Ichthyological Eco-regions of Argentina.* ProBiota - SerieTécnica y Didáctica, La Plata.

Lucena-Moya, P., Duggan, I.C. 2011. Macrophyte architecture affects the abundance and diversity of littoral microfauna. *Aquatic Ecology* 45:279–287.

Marchese, M., Ezcurra de Drago, I., Drago, E. 2002. Benthos invertebrates and physical habitat relationships in the Paraná River flood-plain system. In: McClain, M. (ed.), *The Ecohydrology of South American Rivers and Wetlands.* IAHS Special Publication 6, International Association of Hydrological Sciences (UAHS), pp. 111–132.

Marchese, M.R., Wantzen, K.M., Ezcurra de Drago, I. 2005. Benthic invertebrate assemblages and species diversity patterns of the Upper Paraguay River. *River Research and Application* 21:485–499.

Menni, R.C. 2004. Peces y Ambientes en la Argentina Continental. *Monografías del Museo Argentino de Ciencias Naturales* 5:1–316.

Mesa, L.M., Marchese, M., Montalto, L., Zilli, F. 2012. Bidirectional exchanges of benthic invertebrates in a large river–floodplain system (Paraná River, Argentina). *Annales de Limnologie* 48:425–436.

Mitsch, W.J., Gosselink, J.G. 2009. *Wetland Ecosystems*. John Wiley & Sons, Inc., Hoboken, NJ.

Montoya, J.M., Pimm, S.L., Solé, R.V. 2006. Ecological networks and their fragility. *Nature* 442:259–264.

Moraña, L.B. 2013. Fitoplancton de ríos y arroyos de la Alta Cuenca del río Bermejo, Argentina. *Lhawet* 2:13–20.

Morrone, J.J. 2001. *Biogeografía de América Latina y el Caribe*. Manuales y Tesis. *Revista Sociedad Entomológica Aragonesa (SEA)* 3.

Neiff, J.J. 1990. Ideas para la interpretación ecológica del Paraná. *Interciencia* 15:424–441.

Neiff, J.J. 1996. Large rivers of South America: Towards the new approach. *Verhandlungen des Internationalen Verein Limnologie* 26:167–180.

Neiff, J.J. 2004. *El Iberá…¿en peligro?*. Fundación Vida Silvestre Argentina, Buenos Aires, Argentina, 45 pp.

Neiff, J.J. 2005. Bosques fluviales de la cuenca del Paraná. In: Arturi, M.F., Frangi, J.L., Goya, J.F. (eds.), *Ecología & Manejo de los Bosques de Argentina*. La Plata, Argentina, pp. 1–26.

Neiff, J.J., Iriondo, M.H., Carignan, R. 1994. Large tropical South American wetlands: A review. In: Neimann, R.J., Decamps, H. (eds.), *The Ecology and Management of Aquatic-Terrestrial Ecotones*, United Nations Educational Cultural and Scientific Organization (UNESCO), Washington, D.C., USA, pp. 156–165.

Neiff, J.J., Casco, S., Poi, A.G. 2008. Response of Eichhornia crassipes (Pontederiaceae) to water level fluctuations in two lakes with different connectivity in the Paraná River floodplain. *Revista de Biologia Tropical* 56:613–623 DOI: 10.15517/rbt.v56i2.5612.

Neiff, J.J., Casco, S.L., Mari, E.C.K., Di Rienzo, J.A., Poi, A.S.G. 2014. Do aquatic plant assemblages in the Paraná River change along the river's length? *Aquatic Botany* 114:50–57.

Neiff, J.J., Mendiondo, E.M., Depettris, C.A. 2000. ENSO floods on river ecosystems: Catastrophes or myths?. In: Toenmsnann, F., Koch, M. (eds.), *River Flood Defence, Kassel Reports of Hydraulic Engineering, No. 9., Vol. I, Section F: Flood Risk, Floodplain and Floodplain Management*. Herkules Verlag, Kassel, pp. 141–152.

Nogueira, M.G., Perbiche-Neves, G., Naliato, D.A.O. 2012. Limnology of two contrasting hydroelectric reservoirs (Storage and Run-of-River) in southeast Brazil. In: Samadi-Boroujeni, H. (ed.), *Hydropower-practice and Application*. In-Tech, Croatia, pp. 185–200.

Paggi, J.C. 2004. Importancia de la fauna de "Cladóceros" (Crustácea, Branchiopoda) del litoral fluvial Argentino. *INSUGEO, Miscelánea* 12:239–246.

Pagioro, T.A., Thomaz, S.M. 2002. Longitudinal patterns of sedimentation in a deep, monomitic subtropical reservoir (Itaipu, Brazil- Paraguay). *Archives Hydrobiology* 154:515–528.

Perbiche-Neves, G., Rocha, C.E.F., Nogueira, M.G. 2014. Estimating cyclopoid copepod species richness and geographical distribution (Crustacea) across a large hydrographical basin: Comparing between samples from water column (plankton) and macrophyte stands. *Zoologia* 31:239–244.

Petsch, D.K., Pinha, G.D., Dias, J.D., Takeda, A.M. 2015. Temporal nestedness in Chironomidae and the importance of environmental and spatial factors in species rarity. *Hydrobiologia* 745:181–193.

Poi de Neiff, A.S.G. 2003. Macroinvertebrates living on *Eichhornia azurea* Kunth in the Paraguay River. *Acta Limnologica Brasiliensis* 15:55–63.

Poi de Neiff, A.S.G., Neiff, J.J. 2006. Riqueza de especies y similaridad de los invertebrados que viven en plantas flotantes de la planicie de inundación del Río Paraná (Argentina). *Interciencia* 31:220–225.

Pott, A., Pott, V.J. 2000. *Plantas acuáticas do Pantanal*. EMBRAPA, Brasilia, pp. 1–404.

Pott, A., Pott, V.J. 2004. Features and conservation of the Brazilian Pantanal wetland. *Wetland Ecological Management* 12:547–552.

Pott, A., da Silva, J.S.V. 2016. Terrestrial and aquatic vegetation diversity of the Pantanal Wetland. In: Bergier, I., Assine, M.L. (eds.), *Dynamics of the Pantanal Wetland in South America*. Springer International, Switzerland, pp. 111–151.

Ringuelet, R.A. 1975. Zoogeografía y Ecología de los peces de aguas continentales de la Argentina y consideraciones sobre las áreas ictiológicas de América del Sur. *Ecosur* 2:1–222.

Roberto, M.C., Santana, N.F., Thomaz, S.M. 2009. Limnology in the Upper Paraná River floodplain: Large-scale spatial and temporal patterns, and the influence of reservoirs. *Brazilian Journal of Biology* 69:717–725.

Rodrigues, L.C., Train, S., Bovo-Scomparin, V.M., Jati, S., Borsalli, C.C.J., Marengoni, E. 2009. Interannual variability of phytoplankton in the main rivers of the Upper Paraná River floodplain, Brazil: Influence of upstream reservoirs. *Brazilian Journal of Biology* 69:501–516.

Rojas Molina, F., José de Paggi, S. 2008. Zooplankton in the Paraná River Floodplain (South America) before and after the invasion of *Limnoperna fortunei* (Bivalvia). *Wetlands* 28:695–702.

Schneider, B., Cunha, E.R., Espínola, L., Marchese, M., Thomaz, S.M. 2019. The importance of local environmental, hydrogeomorphological and spatial variables for beta diversity of macrophyte assemblages in a Neotropical floodplain. *Journal of Vegetation Science* 30:269–280.

Schneider, B., Cunha, E.R., Marchese, M., Thomaz, S.M. 2018. Associations between macrophyte life forms and environmental and morphometric factors in a large sub-tropical floodplain. *Frontiers Plant Science* 9:195–210.

Sendacz, S. 2001. Planktonic Copepoda of the Upper Parana River Floodplain lakes (São Paulo and Mato Grosso do Sul, Brazil). *Hydrobiologia* 453/454:367–374.

Souza Filho, E.E. 2009. Evaluation of the Upper Paraná River discharge controlled by reservoirs. *Brazilian Journal of Biology* 69:707–716.

Souza-Franco, G.M., Takeda, A.M. 2000. Invertebrates associated with *Paspalum repens* (Poaceae) at the mouth of Caracu stream (1991–1992), affluent of the Paraná River, Porto Rico (PR) Brazil. *Brazilian Archives of Biology and Technology* 43:317–325.

Stevaux, J.C., Takeda, A.M. 2002. Geomorphological processes related to density and variety of zoobenthic community of the upper Paraná River (Brazil). *Zeitschrift für Geomorphologie* 129:109–129.

Thomaz, S.M., Agostinho, A., Hahn, N.S. (eds.) 2004. *The Upper Paraná River and Its Floodplain*. Backhuys Publisher, Leiden, The Netherlands.

Thomaz, S.M., Bini, L.M., Bozelli, R.L. 2007. Floods increase similarity among aquatic habitats in river-floodplain systems. *Hydrobiologia* 579:1–13.

Torres, M.V., Giri, F., Collins P. 2016 'La Niña' phenomenon and the relationship between decapod populations and fishes in temporarily isolated shallow lakes. *Marine and Freshwater Research* 68:1010–1022 https://doi.org/10.1071/MF16035

Wantzen, K.M., Blettler, M.C.M., Marchese, M.R., Amsler, M., Bacchi, M., Ezcurra de Drago, I.E., Drago, E.E. 2014. Sandy rivers: A review on general ecohydrological patterns of benthic invertebrate assemblages across continents. *International Journal of River Basin Management* 12:163–174.

Wantzen, K.M., Drago, E., da Silva, C.J. 2005. Aquatic habitats of the Upper Paraguay river-floodplain-system and parts of the Pantanal (Brazil). *Ecohydrology and Hydrobiology* 5:107–126.

Wantzen, K.M., Junk, W.J., Rothhaupt, K. O. 2008. An extension of the floodpulse concept (FPC) for lakes. *Hydrobiologia* 613:151–170. DOI 10.1007/s10750-008-9480-3.

Welcomme, R., Halls, A. 2002. Dependence of tropical river fisheries on flow. *Proceedings of the International Large Rivers Symposium*. Scientific Special Publication, pp. 267–283.

Winemiller, K.O. 2004. Floodplain river food webs, generalizations and implications for fisheries management. In: Welcomme, R.L., Petr, T. (eds.), *Proceedings of the Second International Symposium on the Management of Large Rivers for Fisheries Volume 2, Phnom Penh, Cambodia*, Publication of the Regional Alliance for Preservation (RAP) 2004/16, Bangkok, Thailand, pp. 285–309.

Zalocar de Domitrovic, Y. 2005. Biodiversity of phytoplankton in the Paraguay-Paraná river axis. *INSUGEO* 14:229–242.

Zilli, F.L., Marchese, M. 2011. Patterns in macroinvertebrate assemblages at different spatial scales.Implications of hydrological connectivity in a large floodplain river. *Hydrobiología* 663:245–257.

Zilli, F.L., Montalto, L. 2011. Benthic invertebrates in the Middle Paraná River floodplain (Argentina). In: Alvarez, M.A. (ed.) *Floodplains*. Nova Science, New York, NY, USA, pp. 99–126.

6

Climate change impacts

Walter Collischonn, João Paulo Lyra Fialho Brêda, Vinícius Alencar Siqueira, and Rodrigo Cauduro Dias de Paiva

Key messages

- We evaluated long-term (2081–2100) climate change impacts on water resources of Paraná Basin using 25 Global Climate Models (GCMs) and a continental-scale hydrologic-hydrodynamic model.
- Precipitation is projected to increase in the east and south of the basin by up to 10% and decrease in the north and northwest parts of the basin by up to 5%. However, there is still

a considerable disagreement between different GCMs in the magnitude of the changes in precipitation and even in the direction of change (positive or negative) in some regions.

- Rivers that flow from north to south within the Paraná and Paraguay headwaters, including the western rivers draining the Andes, are projected to face discharge reductions from 5–30%. The Paraguay river may experience a larger decrease during the low flow period. There is no agreement, however, about the future changes in the discharge of the main Paraná River itself. As a result of increased potential evapotranspiration and reduced precipitation, the northern part of the Paraná River Basin is projected to become drier, as expressed by changes of up to 0.5 in the aridity index.
- Reduced discharges in the Paraguay River and the Upper Paraná (especially in the Paranaíba tributary) may cause negative impacts on navigation and energy production. Other activities like agriculture and livestock production can be severely affected by the potential drier conditions.

Introduction

Lying in the central part of South America, the Paraná River Basin is one of the largest hydrological systems of the world (~2.6 million km^2), draining waters from many countries, including Brazil, Paraguay, Argentina and Bolivia. The eastern part of the basin, mostly located in Brazil, receives more rainfall than the western part. The upper Paraná River Basin receives 1,300 to 2,000 $mm.y^{-1}$, while the Chaco region, located west of the Paraguay River, receives precipitation as low as 300 $mm.y^{-1}$ (Berbery and Barros, 2002). The precipitation patterns and influences on geomorphology in the basin are reviewed in Chapter 3 of this book. Because of these differences in rainfall, the Paraná and its major tributary, the Paraguay River, are very dissimilar in terms of discharge. The mean discharge of the Paraguay River close to its confluence with the Paraná is around 3,900 m^3 s^{-1}, while the Paraná itself discharges circa 12,000 $m^3.s^{-1}$ just upstream of the confluence, despite having a slightly smaller drainage area (Collischonn et al., 2016).

An important feature of the Paraná River and some of its tributaries is the presence of low frequency variability (e.g., longer than a decadal time scale) and non-stationary behaviour of hydroclimatic variables, with relatively long sequences of above or below average rainfall and streamflow (García and Vargas, 1996, 1998; Genta et al., 1998; Tucci and Clarke, 1998; Collischonn et al., 2001; Krepper et al., 2008). This feature was noted very early by Charles Darwin in his field notes taken during the Voyage of the Beagle. Changes in trend signals regarding the basin hydrology have been the subject of research for at least two decades, and several causes have been suggested, ranging from interannual variability in precipitation (Collischonn et al., 2001), land use alterations (Doyle and Barros, 2011; Lee et al., 2018), solar activity (Antico and Kröhling, 2011) and the frequency of El Niño events (Boulanger et al., 2005; Krepper and Venturini, 2009).

Another physical process that may be causing, at least partially, the non-stationary behaviour of the hydrological variables in this region is global climate change (Barros et al., 2006). On a global basis, climate projections indicate that a temperature rise above 1.5°C is expected by the end of this century, with changes in precipitation patterns around the globe (Flato et al., 2013). In South America, the annual precipitation mean is likely to increase in the southeast but decrease in most areas between latitudes 5 to 20 degrees south (Chou et al., 2012). These climate trends may affect hydrological events and as a consequence, influence socio-economic and environmental conditions over the entire continent (Reyer et al., 2017). Predicted water related impacts in South America include reduced water supply over the long term due to glacier melt in the Andes (Chevallier et al., 2011; Vuille et al., 2018), a decrease in hydroelectric production (Silveira et al., 2019), modifications on biogeochemical properties of the Amazon floodplains (Melack and Coe, 2012), increased flood risk in western Amazonia and southeast South America due to extreme events (Marengo et al., 2009), and increased risks of desertification in northeast Brazil (Marengo and Bernasconi, 2015).

Within the Paraná River Basin, climate change effects on hydrology can have impacts on the environment and on human water uses. The basin is home to nearly 90 million people (CIESIN, 2018) who use water for domestic and industrial purposes. Water resources are also used extensively for agriculture, with continuously growing irrigated areas. The basin is particularly notable for the use of water to produce energy through hydroelectric power plants. Itaipu, the world's second largest hydropower plant, is located on the Paraná River, just upstream of the confluence with the Iguazu River. Downstream of Itaipu, the Yaciretá hydropower plant located between Argentina and Paraguay is one of the most important electric power sources for these two countries. Navigation is also a major activity, with important fluvial ports in all countries. The Paraná River is navigable along most of its route in Argentina and Paraguay, and the Paraguay River is also used for transport from its mouth up to the city of Cáceres in Brazil. The rivers Tietê and Paraná are also navigable in Brazil, upstream of the Itaipu dam. Water management challenges, such as dealing with extreme weather events like floods and droughts are pressing issues in the Paraná River Basin. Floods, for instance, have impacted several urban centers along the Paraná and its tributaries in the past, notably the cities of Asuncíon (Paraguay) and Santa Fé (Argentina), and will probably do so again in the coming decades.

In this chapter, we evaluate the long-term (2081–2100) climate change impacts on the water resources of the Paraná River Basin by using the predictions of 25 Global Climate Models (GCMs) used in the IPCC AR5 report and a continental-scale, hydrologic-hydrodynamic model validated for this region. We provide insights on where changes in the water budget and river discharge are expected and how much these changes are significant.

Models and scenarios

Projections of climate change impacts on water resources are commonly investigated using hydrological and climatological models. GCMs simulate time series of several hydrometeorological variables through the Earth, such as temperature, precipitation, atmospheric moisture and pressure. These models are run over a long time period that characterizes the past and current climate, and projects the climate for the future. The difference between current/past and future time series of atmospheric variables indicates the expected climatic changes. Surface hydrometeorological variables from the GCMs are used as inputs for the hydrological models over these two distinct periods in order to simulate hydrological variables such as streamflow. Therefore, streamflow can also be compared between current and future climate scenarios.

However, there are several sources of uncertainty in this kind of assessment. One of the most important sources of uncertainty is the choice of the GCM (Nóbrega et al., 2011). The accuracy of the GCM is limited by the current degree of understanding of the physical processes and by numerical/mathematical simplifications. The performance of GCMs has been improving over time (Flato et al., 2013) but there are still at present significant uncertainties concerning the variables that are important for hydrological modeling, such as precipitation. Due to such uncertainty, it is usually recommended that projections be made based on ensembles of different GCMs, instead of relying excessively on one single GCM. For instance, in

South America, Gulizia and Camilloni (2015) demonstrated that, in general, the mean precipitation estimated from several GCMs yields lower errors than the simulated precipitation using a single GCM. Therefore, to assess climate change impacts on water resources in the Paraná River Basin, we used the outputs from 25 GCMs and applied them as inputs to a continental-scale hydrological model.

The Paraná hydrological model

To assess the possible impacts of climate change on the Paraná River Basin, we applied the large-scale hydrological model, MGB (acronym for *Modelo de Grandes Bacias*, or *Large Basins Model*), as described by Collischonn et al. (2007). This is a large-scale distributed hydrologic hydrodynamic model that was tested several times in South American river basins and is being constantly adapted to improve its capability to represent large basins, notably in regions with mild slopes and large floodplains (Paiva et al., 2011, 2013; Pontes et al., 2017; Fleischmann et al., 2018). The model has been used before to assess climate change impacts (Lima et al., 2014;), including regional studies in parts of the Paraná River Basin, such as the Grande River (Nóbrega et al., 2011) and the Upper Paraguay in the Pantanal region of Brazil (Bravo et al., 2014).

MGB is a conceptual distributed model, in which the river basin area is divided into smaller catchments, and into Hydrological Response Units (HRU) defined by land use, land cover, geological features and soil types. Evapotranspiration is calculated through the Penman-Monteith equation and leaf area index is used to infer vegetation interception capacity. Water balance is calculated for each HRU within each catchment, and streamflow is routed within the catchment to the river network by simple linear reservoirs. Flow routing in river channels is computed using a hydrodynamic model (Pontes et al., 2017), accounting for inundation and infiltration from flooded areas into the unsaturated soil column in floodplains (Fleischmann et al., 2018). These features are essential to simulate basins with large flat regions. Here, we applied the MGB model that was developed for the whole South American continent by Siqueira et al. (2018). This model has been validated with hundreds of *in situ* gauges and remote sensing datasets and it has compared favorably to Global Hydrological Models with respect to predicting river discharges. The model also demonstrated good performances for predicting historic water storage, evapotranspiration and river water levels, which increases confidence in the model for climatic assessments.

Climate change scenarios and GCMs

The GCM data used in this study is a subset of the CMIP5 multi-model ensemble snapshot of March 15, 2013, which was discussed in the IPCC Working Group I AR5 report. This dataset consists of monthly means of climate variables from 1850 to 2100. Our assessment was conducted considering two distinct periods of 20 years: (i) a baseline period from 1986–2005, and (ii) a future period from 2081–2100. We selected the GCMs listed in Table 6.1, that provide the prescribed MGB input variables, simulation periods and Green House Gas (GHG) emission scenarios.

GCMs are driven by GHG scenarios to simulate land, ocean and atmosphere interactions and, consequently, alterations in climate. There are four scenarios (RCP 2.6, 4.5, 6.0 and 8.5) on the fifth phase of the Coupled Models Inter-comparison Project (CMIP5), that account for different projections of industrial development and environmental policies, which

Table 6.1 List of GCMs used to assess the impacts of climate change on the water resources of the Paraná River Basin.

Model	*Institution*	*Country*	*Resolution of precipitation grid (degrees)*	
			Longitude	*Latitude*
ACCESS1.0	Commonwealth Scientific and Industrial Research Organisation/ Bureau of Meteorology (CSIRO-BOM)	Australia	1.25	1.875
ACCESS1.3			1.25	1.875
BCC-CSM1.1	Beijing Climate Center (BCC)	China	2.7906	2.8125
BCC-CSM1.1 (m)			1.1215	1.125
BNU-ESM	Beijing Normal University (BNU)		2.7906	2.8125
CanESM2	Canadian Centre for Climate Modelling and Analysis (CCCma)	Canada	2.7906	2.8125
CNRM-CM5	Centre National de Recherches Météorologiques (CNRM-CERFACS)	France	1.4008	1.40625
CSIRO-Mk3-6-0	Commonwealth Scientific and Industrial Research Organisation (CSIRO)	Australia	1.8653	1.875
GFDL-CM3	Geophysical Fluid Dynamics Laboratory (GFDL)	USA	2	2.5
GFDL-ESM2G			2.0225	2.5
GFDL-ESM2M			2.0225	2.5
GISS-E2-H	NASA Goddard Institute for Space Studies (NASA-GISS)		2	2.5
GISS-E2-R			2	2.5
HadGEM2-CC	Met Office Hadley Centre (MOHC)	United Kingdom	1.25	1.875
HadGEM2-ES			1.25	1.875
HadGEM2-AO	MOHC + National Institute of Meteorological Research, Korea Meteorological Administration (NIMR-KMA)	UK + South Korea	1.25	1.875
INM-CM4	Russian Academy of Sciences, Institute of Numerical Mathematics (INM)	Russia	1.5	2
IPSL-CM5A-LR	Institut Pierre Simon Laplace (IPSL)	France	1.8947	3.75
IPSL-CM5A-MR			1.2676	2.5
IPSL-CM5B-LR			1.8947	3.75
MIROC-ESM-CHEM	Atmosphere and Ocean Research Institute (The University of Tokyo), National Institute for Environmental Studies and Japan Agency for Marine-Earth Science and Technology (MIROC)	Japan	2.7906	2.8125
MIROC-ESM			2.7906	2.8125
MIROC5			1.4008	1.40625
MRI-CGCM3	Meteorological Research Institute (MRI)		1.12148	1.125
NorESM1-M	Bjerknes Centre for Climate Research, Norwegian Meteorological Institute (NCC)	Norway	1.8947	2.5

directly affect GHG emissions. The acronym RCP stands for Representative Concentration Pathways, which indicate increases of radiative forcing in $W.m^{-2}$ by the end of this century compared to the pre-industrial era, mainly due to GHG concentrations in the atmosphere. For instance, the RCP 2.6 represents a mitigated scenario where the GHG emissions are significantly reduced over time, while the RCP 8.5 stands for an increasing emissions scenario that reaches high GHG concentrations in the atmosphere. We investigated the RCP 4.5 scenario (Thomson et al., 2011), which represents a moderate scenario that has been commonly tested in the literature.

We used 25 GCMs (Table 6.1) under the RCP 4.5 scenario to simulate climate change impacts. Each GCM yielded atmospheric outputs from years 1850 to 2005 as the 'historical period' and from 2006 to 2100 to cover the RCP simulations. These 25 GCMs were selected from a subset of the CMIP5 multi-model ensemble snapshot of March 15, 2013. The subset is discussed on the Fifth Assessment Report – Physical Basis, conducted by the IPCC Working Group I (Stocker et al., 2013). Two distinct 20-year periods from the GCM simulations were selected to estimate changes in climate: the baseline (1986–2005) and future (2081–2100) periods. The differences between these two periods represent the climatic changes that are expected to impact water resources in the Paraná River Basin by the end of this century.

GCMs often present systematic errors that have significant effects on hydrological results and such biases should be removed from the analyses. In this study, the bias correction was conducted through the delta change approach, which consists of comparing GCM results from the baseline and future periods and applying this relative change on observed databases to simulate an unbiased future. Thus, the atmospheric outputs of the GCMs were not directly used as input on the hydrological model, but rather the simulated difference between 'present' and 'future' simulations for each GCM, building an ensemble of 25 impact projections. This ensemble was compared to a reference simulation driven by observed databases, namely the MSWEP v1.1 daily precipitation (Beck et al., 2017), extracted from 1990 to 2010, and the CRU v.2 long-term monthly means of surface meteorological variables, the latter referring to the 1960–1990 period.

Climate change impacts on water resources

Projected changes in precipitation

Figure 6.1 shows projected changes in mean annual precipitation within the Paraná River Basin, computed as the mean change of the 25 GCMs listed in Table 6.1. These simulations show that precipitation is projected to decrease in the northern part of the basin, and to increase in the southeast part of the basin. Precipitation reduction is in the range of 2–5%, as shown in the region colored orange in Figure 6.1, while the precipitation increase is in the range of 2–10%, as shown in the region in blue in Figure 6.1.

The headwaters of the Paraguay River close to Cuiabá (Figure 6.1) are in the middle of a region where precipitation is projected to decrease, while the headwaters of the Paraná, between São Paulo and Brasília are in a transition zone, where projected precipitation changes vary from a decrease in the north to an increase in the south. The largest projected increase in precipitation can be seen in the headwaters of the Iguaçu River to the east of Foz do Iguaçu and along the main course of the Paraná River itself, down to Rosario, especially in the region to the east of the river (Figure 6.1). A decrease in precipitation is also projected in the western extreme of the basin, along the Andes, at the headwaters of the Salado and Pilcomayo rivers.

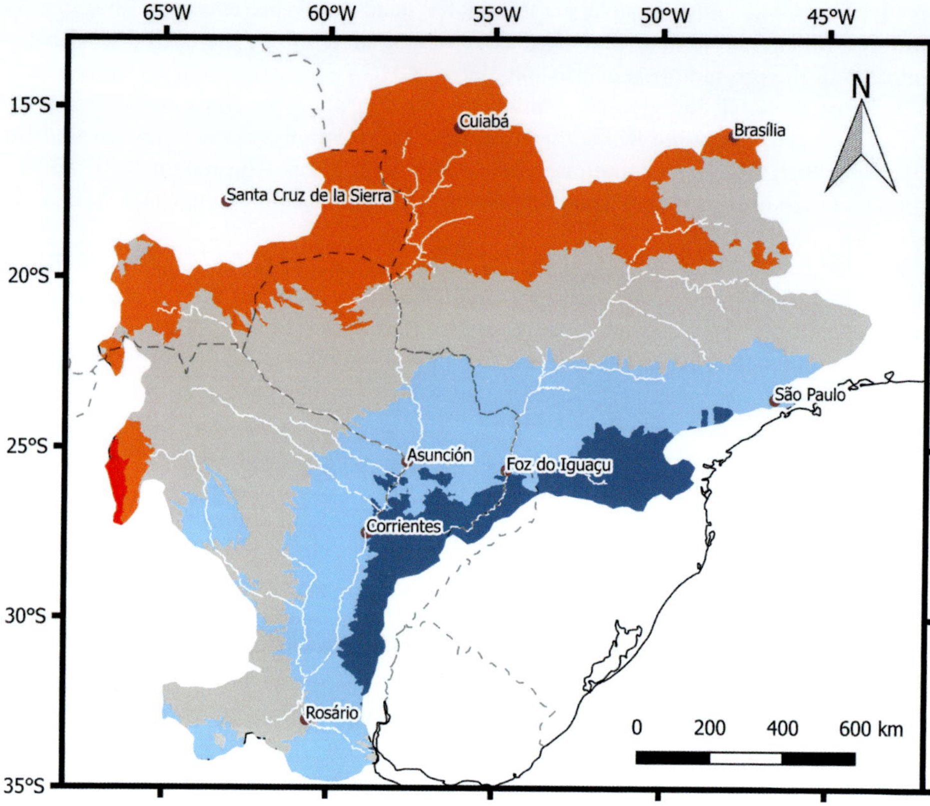

Figure 6.1 Projected change in mean annual precipitation in the Paraná River Basin (average of 25 GCMs). Blue (red) colours indicate precipitation increase (decrease).

Projected changes in actual evapotranspiration

Figure 6.2 shows the projected changes in evapotranspiration within the Paraná River Basin, using outputs from the hydrological model with input data from the 25 GCMs listed in Table 6.1. These simulations show that evapotranspiration is projected to increase over most of the basin, most notably in the east/southeast part of the basin. This increase is related to higher air temperatures and more rainfall in this region. In the northern part of the Paraná basin, on the other hand, projections do not show changes in actual evapotranspiration. This is a result of the combined effects of increased temperature, which drive an increase in potential evapotranspiration, with reduced precipitation, which drives a reduction in water availability and, thus, actual evapotranspiration.

Projected changes in runoff

Figure 6.3 illustrates the projected changes in annual local runoff within the Paraná River Basin, using outputs from the hydrological model, with input data from the 25 GCMs listed in Table 6.1. These simulations show that runoff is projected to decrease in the northern part of the Paraná River Basin, most notably in the Upper Paraguay river and the headwaters of the

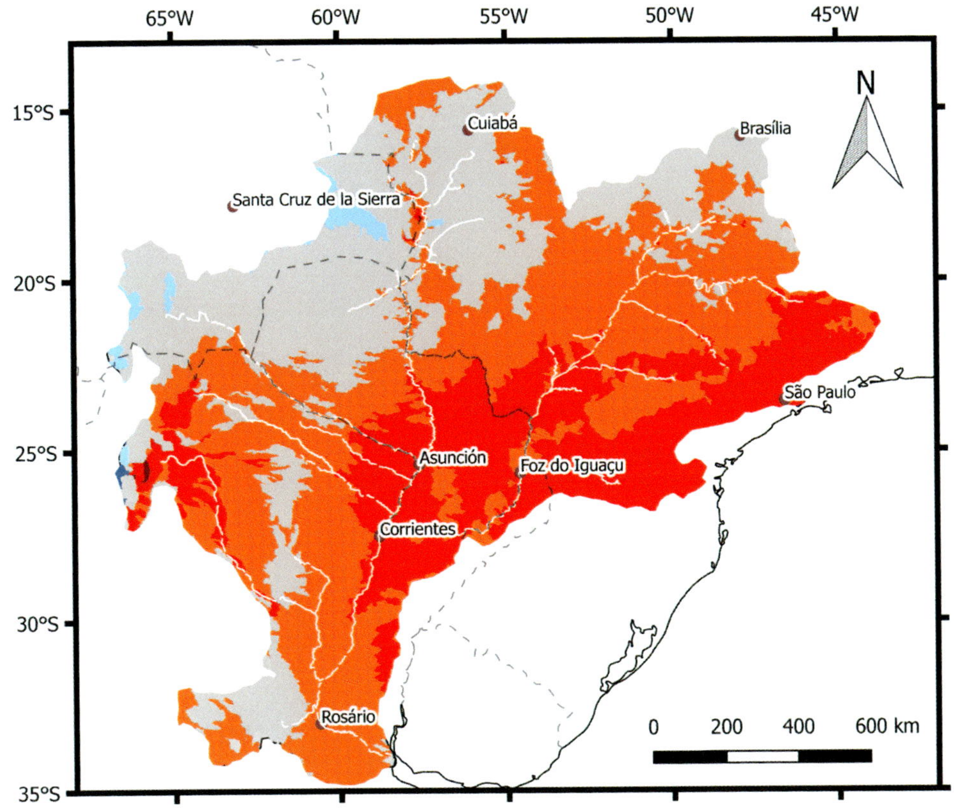

Figure 6.2 Projected change in annual actual evapotranspiration in the Paraná River Basin (mean of 25 GCMs). Blue (red) colors indicate evapotranspiration decrease (increase).

Pilcomayo, Bermejo and Salado rivers, located in the northwestern part of the basin. Runoff may decrease from 10–30% in some parts of this region.

Conversely, an increase in runoff is projected to occur in the southeastern part of the basin, to the east of the lower reach of the Paraná, downstream of Corrientes, and in the Iguaçu River Basin. Runoff may increase by up to 30% in some parts of this region (Figure 6.3). For a large part of the basin, in the region from São Paulo to Asunción, and from there to Argentina west of the Paraná, small changes in runoff are projected under the climate change scenarios, as shown in grey in Figure 6.3. It should be noted that the projected changes in runoff are larger than the projected changes in precipitation, due to the additive effect of precipitation and evapotranspiration changes that are also amplified over runoff.

Projected changes in river discharge

While changes in runoff are related to the amount of water that is produced at a particular location, it does not provide information about the spatially integrated impacts of climate change in large river basins, where water is conveyed from one region to the other, over thousands of kilometers, through an extensive drainage system. Projected changes in river discharge, on

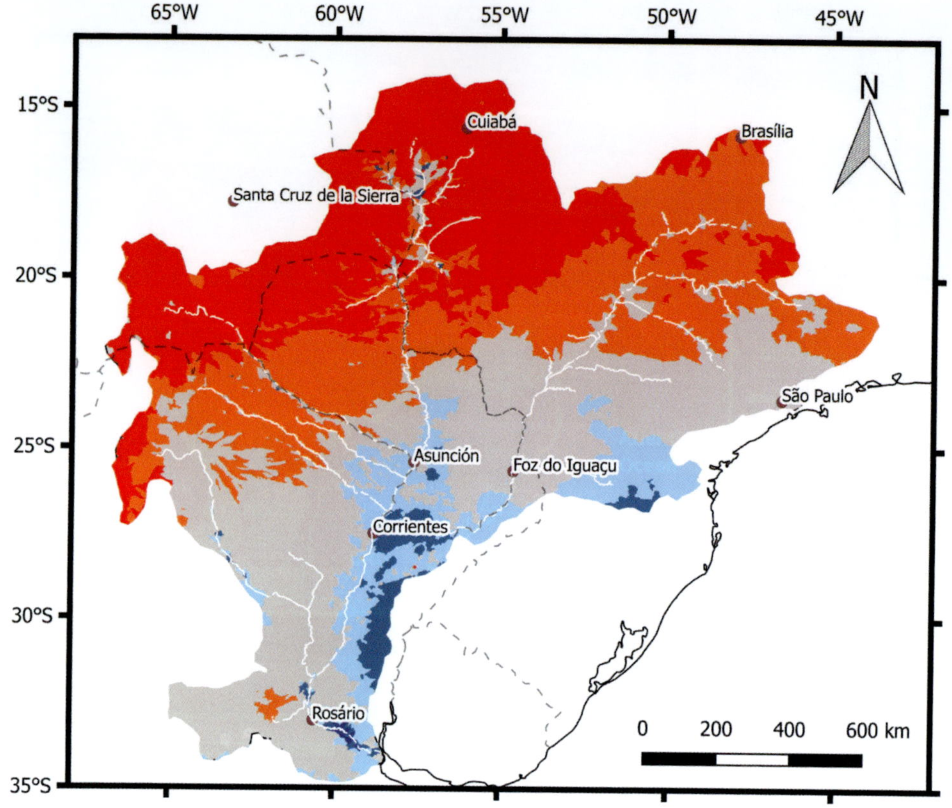

Figure 6.3 Projected change in annual local runoff generation in the Paraná River Basin (mean of 25 GCMs). Blue (red) colors indicate runoff increase (decrease).

the other hand, show the integrated impact of changes in rainfall, evapotranspiration and water balance over the whole contributing area of a river reach. These projections are shown in Figure 6.4, which shows that both the Upper Paraguay and Upper Paraná are projected to have a reduction in discharge. This reduction can be as high as 30% in some places, according to the mean of 25 GCM outputs. Changes are projected to be more intense in the Upper Paraguay than in the Upper Paraná.

Projections indicate that the two primary tributaries of the Paraná, namely the Paranaiba (South of Brasilia) and the Grande (north of São Paulo) may have 5–10% reductions in average annual streamflow. As a result, the Paraná itself shows a projected reduction in discharge of the same order of magnitude downstream to Guaíra, close to Foz do Iguaçu (Figure 6.4). However, due to the contribution of tributaries to the east of the river, such as the Tietê, Paranapanema and Iguaçu, that show slight increases or no projected change in discharge at all, the Paraná shows no projected change in discharge in the region of Foz do Iguaçu downstream to Corrientes, and further downstream to Rosario. Rivers flowing from the Western part of the basin, down the Andes and pre-Andes mountains, including the Pilcomayo, Bermejo and Salado rivers are projected to have a decrease in streamflow. However, such decrease is projected to weaken in the downstream direction, as a result of increased rainfall along the Paraná main course.

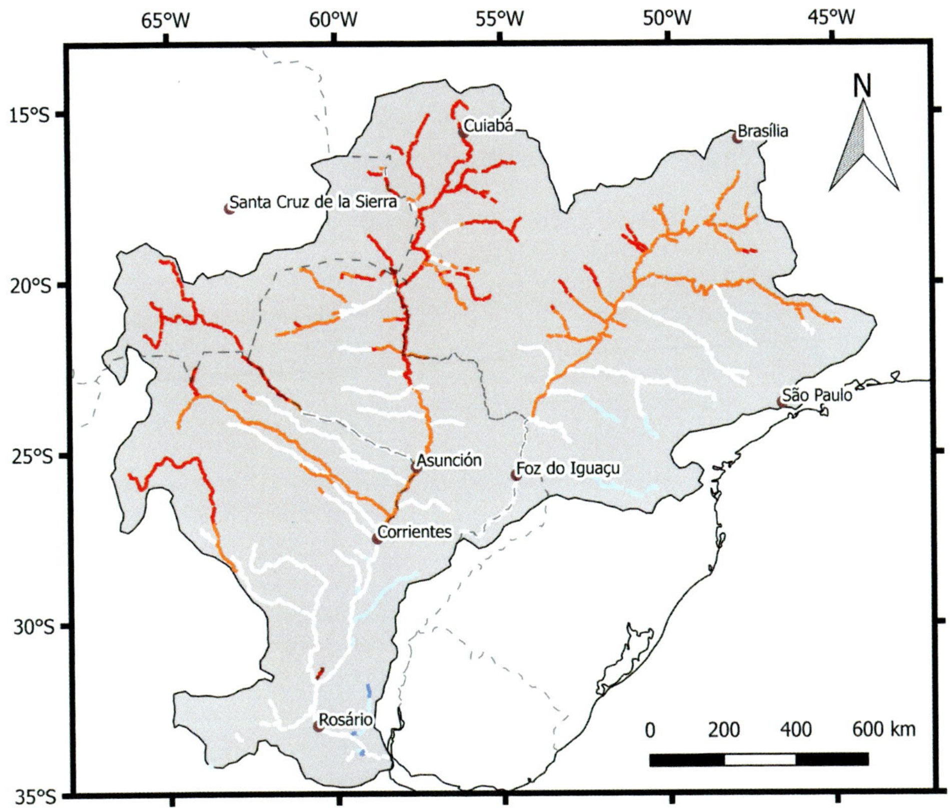

Figure 6.4 Projected change in mean streamflow in the Paraná River Basin (mean of 25 GCMs). Blue (red) colors indicate streamflow increase (decrease).

Figure 6.5 shows hydrographs of mean monthly streamflow climatology, derived from the hydrological model simulation, at three selected locations: (i) the Paraguay River at Asuncion, (ii) the Paraná River at Foz do Iguaçu, upstream of its confluence with the Paraguay, and (iii) the Paraná River at Rosario, close to its mouth at the estuary of the Rio de la Plata. This figure shows the reference simulation (solid black line) as the resulting discharge, considering the reference climate condition (current climate), while the grey band represents a range of future climate scenarios (interquartile range).

From Figure 6.5a, it can be observed that the discharge of the Paraguay River at Asunción is projected to decrease over the annual average but shows a larger decrease during the low flow period (i.e. July to October). No clear changes are projected for the Paraná River at Foz do Iguaçu and Rosario in most of the year (Figure 6.5b, 6.5c) but during the low flow season (July to October) there is a clearer pattern of reduced discharge.

Projected changes in aridity

An aridity index (AI) is a metric to express the degree of dryness of the climate of a region or of a certain location. This index serves to identify regions that suffer from a deficit of available water, a condition that can severely affect activities such as agriculture and stock-farming.

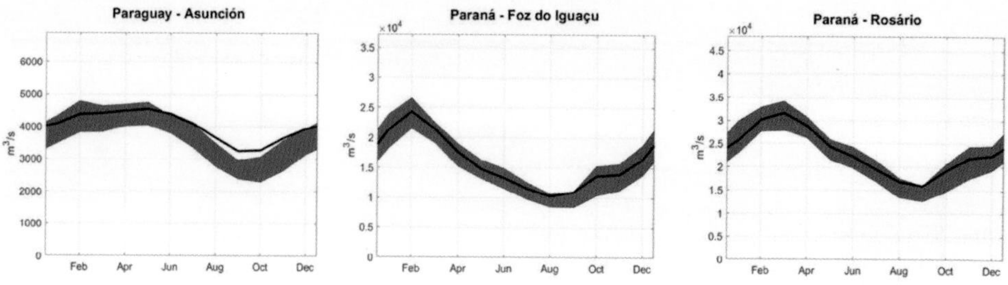

Figure 6.5 Projected change in streamflow ($m^3.s^{-1}$) over an annual hydrological cycle at three reference locations in the Paraná River Basin: (a) Paraguay River at Asuncion; (b) Paraná River at Foz do Iguaçu; (c) Paraná River at Rosario. The solid black line represents the reference simulation and the grey band represents the interquartile range of future scenarios.

Following the approach described by Yang et al. (2018), the Aridity Index was calculated by the ratio of Potential Evapotranspiration to Precipitation (PET/Prec) for the reference (current) and projected (future) climate, for each one of the 25 GCMs. Potential evapotranspiration was calculated using the Penman-Monteith equation within the MGB hydrological model. Relative changes in the Aridity Index were then derived from the current and future climate calculated values, and the average of this change, over the 25 GCMs was then obtained, resulting in the map illustrated in Figure 6.6. Regions colored in orange, and specially in red and dark red, are projected to become more arid in the future.

The predictions illustrated in Figure 6.6 indicate that the northern part of the basin may become more arid, including parts of the Chaco region which extends through western Paraguay and northern Argentina. Some of those regions to the west are semi-arid to arid even now. On the other hand, the region east of the lower reach of the Paraná, between Corrientes and Rosario, is projected to become less arid, as shown by the blue region in Figure 6.6.

Agreement between GCMs

There is high uncertainty in making projections of future climate conditions in a region, such as the Paraná River Basin, and this is mostly due to differences in the GCMs used to produce them. These uncertainties are transferred and even amplified when these climate projections are used as input to hydrological models to address potential changes in variables such as streamflow and evapotranspiration. As a result, projections of changes in river discharge due to climate change may differ greatly in magnitude at a single location, depending on the GCM that was used to generate the input to the hydrological model. There may even be disagreement in terms of the predicted directions of change, with some of the projections suggesting an increase and other projections suggesting a decrease in river discharge.

As an approach to assess the uncertainty of projected changes in a given hydrological variable, we identified regions where at least 2/3 of the projections (i.e., 17 out of the 25 GCMs) agreed on the sign (positive or negative) of change in both precipitation (given by the GCMs) and river discharge (calculated by the hydrological model). The result is shown as a map of the Paraná River Basin in Figure 6.7. Regions where there is agreement that future precipitation will be lower than in the current climate scenario are shown in orange in Figure 6.7, while blue is used to show regions where there is agreement that precipitation will increase in the future. In the region shaded in grey there is no agreement between the GCMs on the sign of

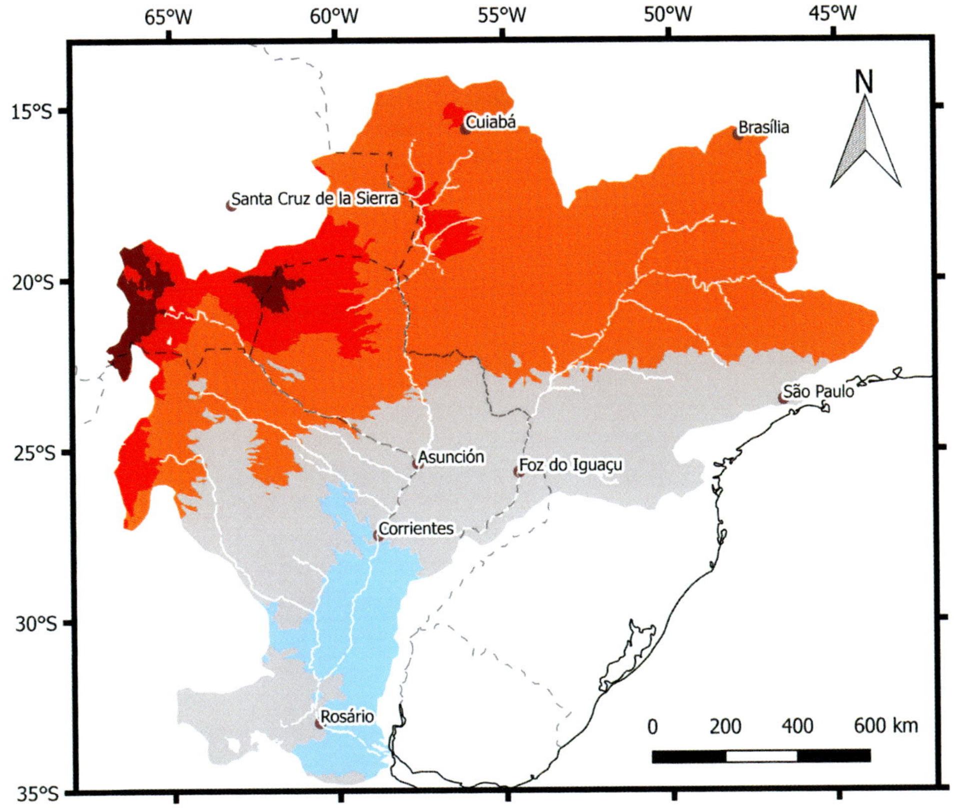

Figure 6.6 Projected change in aridity index, given by the ratio of Potential Evapotranspiration to Precipitation (PET/Prec >5.0 is arid; from 2.0 to 5.0 is semi-arid; from 1.5 to 2.0 is dry sub-humid and less than 1.5 is humid).

change in precipitation. Following the same scheme, river reaches shown in red in Figure 6.7 are the ones where at least 17 out of the 25 results of the hydrological model agree that future river discharge will be lower than discharge calculated considering current climate conditions. River reaches shown in white in Figure 6.7 are those where there is no agreement about the direction of change. Blue shows river reaches with agreement of increased discharge, but none of the rivers showed this kind of result at the spatial scale used to generate Figure 6.7.

These analyses show that there is agreement among the GCMs that the Upper Paraguay River sub-basin between Santa Cruz de La Sierra in Bolivia and Cuiabá in Brazil will receive less rainfall in the future (Figure 6.7). The same applies to the regions in the headwaters of the Pilcomayo and Salado rivers. On the eastern and southern parts of the basin, there are some regions in which there is agreement between models that future precipitation will be higher than the current levels. However, in most of the river basin there is no agreement whether future precipitation will be lower or higher relative to what it is now (i.e. the region shown in grey in Figure 6.7). Despite the lack of agreement among the models on the sign of change in precipitation, there is agreement in the predictions that the main rivers flowing from the north within the Upper Paraná will experience a decrease in river discharge as a result of climate change (i.e. river reaches shown in red in Figure 6.7). In the Paraguay River, there is

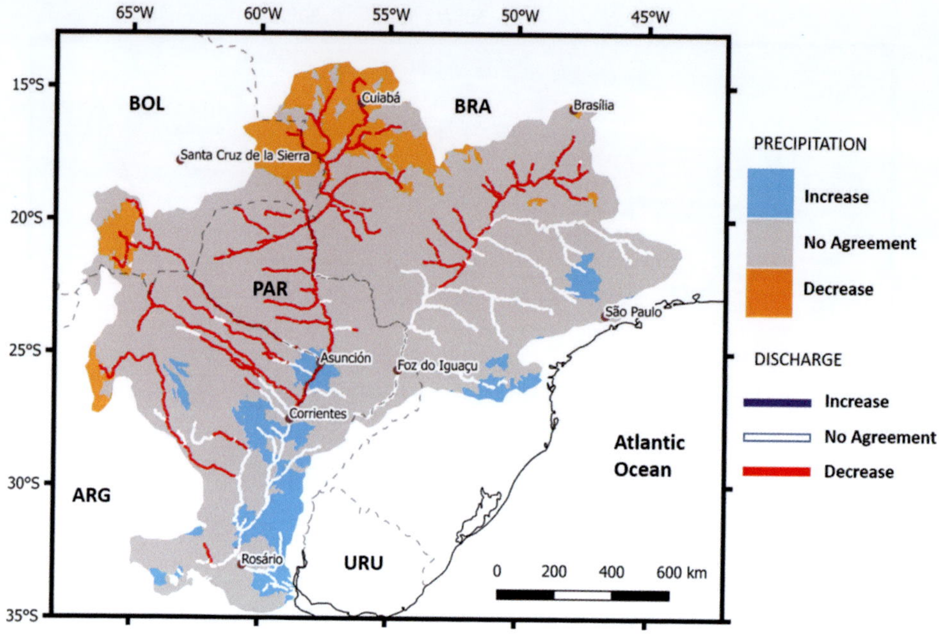

Figure 6.7 Projected changes in mean streamflow (lines) and average annual rainfall (areas) in the Paraná River Basin. Orange and red indicate GCMs agreement for decreased rainfall or streamflow, while grey and blue indicate disagreement and agreement for increase, respectively.

agreement with the predictions that future streamflow will be lower from its headwaters around Cuiabá down to its mouth at the confluence with the Paraná River, close to Corrientes in Argentina (Figure 6.7).

In the Upper Paraná Sub-basin, there is no agreement about the direction of change in streamflow of the main rivers coming from the east, including the rivers Grande, Tietê and Paranapanema (near São Paulo). On the Paranaíba, on the other hand, with headwaters close to Brasilia, there is agreement that discharge will be lower in the future than it is under the current climate conditions (Figure 6.7). This prediction that streamflow will be reduced continues downstream of where the rivers Paranaíba and Grande join to form the Paraná and can still be observed down to the confluence of the Paraná with the Paranapanema. From that point down to the mouth of the Paraná River, close to Rosario, there is no agreement among the models whether river discharge will increase or decrease under future climate conditions.

Agreement with climate projections from other studies

The results presented in the previous sections, obtained by running a continental-scale hydrologic-hydrodynamic model (Siqueira et al., 2018) with climate projections from 25 different GCMs of the CMIP5 are in agreement with some previous studies. Nóbrega et al. (2011) analyzed potential changes of river discharge of the Rio Grande (one of the tributaries of the Paraná in Brazil) and found that under a scenario with a rise in global mean air temperature of 2°C, projected changes in mean river flow ranged from −20% to +18%, depending on the GCM used. This result agrees with the predictions presented here, for which the Rio Grande has a mean projection of decreasing streamflow, however there is no agreement between the GCMs about the direction of the change.

Amorim and Chaffe (2019) reviewed a collection of 42 peer-reviewed articles about climate change impacts on water resources in Brazil and summarized the main conclusions by region. For the Paranaíba River Basin, the four studies reviewed all concluded that future climate will bring a reduction in mean river discharge. This is consistent with the results described here. However, in the present study, we showed that within the part of the Paraná River Basin that lies within Brazil, there is a gradient of projected change, with reduced streamflow in the north and increased streamflow in the south.

On the other hand, the predictions presented here do not agree with results presented previously by Bravo et al. (2014) and Montroull et al. (2018) on projected climate change impacts on streamflow of the Paraguay River. Bravo et al. (2014) found that nearly half of projections predicted an increase in river discharge in the future, and the other half predicted a decrease in river discharge, while Montroull et al. (2018) projected a slight increase in river discharge. In the present study, we showed that there is agreement among the different GCMs that the Paraguay River basin will experience a decrease in streamflow under future climate conditions. This inconsistency may be because the data outputs of a new generation of GCMs were used in the present study. Another possible explanation is related to differences in the hydrological model. The hydrological model used by Bravo et al. (2014) was a combination of MGB (Collischonn et al., 2007) and River Analysis System of the Hydrologic Engineering Centre (US Army Corps of Engineers), and was detailed in terms of the simulation of river hydraulics but did not account for evapotranspiration in large floodplain areas (Bravo et al., 2011). The hydrological model applied by Montroull et al. (2018) was also simplified in the representation of floodable areas. In the present study, we applied a new coupled hydrology–hydrodynamic version of the MGB model (Pontes et al., 2017), which was further enhanced to take into account re-infiltration of water and evapotranspiration in floodplain soils (Fleischmann et al., 2018).

It should also be highlighted that the assessment of climate change impacts shown in this chapter did not consider the impacts on rivers driven by flow from snow melt. Some of the rivers flowing from the Andes are partly fed by snowmelt and, as pointed out by Barros et al. (2015), if mountain temperatures rise steadily, as projected by GCMs, snow will melt earlier in spring, increasing streamflow during the early spring and reducing it in summer, precisely at the time of increased irrigation demand for agriculture.

Potential socio-economic impacts of climate change

Impacts on agriculture

In the basins of the Upper Paraná and Upper Paraguay, the land is extensively used for producing soybeans, sugar cane, cotton, coffee, oranges, maize and other crops. Most of this production is currently not irrigated. An increase in the Aridity Index, as projected for the northern part of the basin, will possibly have an impact on agriculture and its water demand, since the demand for irrigation of crops will probably increase. Increased water abstractions for irrigation will impact other water uses, like hydropower production and navigation. The projected increase in the Aridity Index that we predict here for the northern part of the basin is in agreement with results shown by Sordo-Ward et al. (2017) who suggested that this region will face an increase in droughts.

Barros et al. (2015) state that if future climate changes in northern Argentina will bring warmer and wetter conditions, this may favor agriculture in new areas to the west and to the south, and it may even be possible to grow new crops, such as coffee. In Argentina, climate changes observed over the past few decades have favored the yields of the main crops, mainly due to an increase in rainfall in relatively dry regions. However, it is uncertain if these changes will be sustained in the future, and if the gains in agricultural productivity will not be counterbalanced by losses due to extreme climatic events, such as intense rainfall events (Barros et al., 2015). Zullo et al. (2011) also suggested that there may be potential benefits of regional warming on agriculture in Brazil. However, Assad et al. (2019) state that, for the agriculture in the country as a whole, negative impacts resulting from a temperature increase will outweigh positive effects.

Impacts on hydropower

The Paraná River Basin is one of the basins in the world with the greatest level of hydropower production, with hundreds of dams, many of them quite large, as described in Chapter 4 in this book (Tundisi and Matsumura-Tundisi, in press). More than half of the hydropower production in Brazil and 100% in Paraguay comes from dams located on rivers in the basin. Some of the largest electric power plants of Argentina are also within the basin. Therefore, one of the most important concerns related to climate change in the Paraná River Basin is the implications for electric power generation.

Our results show that a reduction can be expected in hydropower production in the region because some of the rivers flowing from the north, notably the Paranaíba, are projected to have lower river discharge under future climate conditions. There is even relatively strong agreement between different GCMs on this reduction in streamflow. The Paranaíba and its tributaries are heavily used for hydropower production, with more than 7,000 MW of installed capacity. A potential reduction in hydropower production in the region was also predicted earlier by Pereira et al. (2009), who evaluated the impacts of climate change on the Brazilian renewable energy sector. More recently, an assessment conducted by Silveira et al. (2019) also concluded that the Brazilian part of the Paraná River Basin that is very important for hydropower production, is projected to have decreased rainfall in the future.

However, based on results shown here, it seems that this reduction in hydropower production will be more important in Brazil. There is no agreement on the sign of predicted changes in discharge within the Paraná River along the reaches that are used for hydropower production in Paraguay and Argentina. These results are also in general agreement with previous results by Lima et al. (2014), Popescu et al. (2014) and Queiroz et al. (2019) who projected hydropower production potential decrease in the northern parts of the basin and increase in the central part of the basin.

Impacts on navigation

Climate change may affect navigation in rivers by altering water depth along the waterways. The change in depth may be due to reduced river discharge or due to changes in river morphology. Along the Lower Paraná, within Argentina, our results suggest that navigation will not be affected, since the river discharge is not projected to change in the future as a consequence of climate change. This is partially in agreement with a previous study by Guerrero et al. (2013), who used discharge outputs from a hydrological model run using climate projections from four GCMs to investigate the effect of climate change on the dredging cost required to maintain the navigation channels in the Paraná waterway, and concluded that dredging costs will decrease due to increased river discharge.

Along the Paraguay River, our results suggest that navigation will be negatively impacted, due to a reduction in river discharge, particularly during the low flow months. The Paraguay River needs to be continuously dredged, especially in the northern reach and therefore, a future reduction in river discharge will probably lead to increased costs of dredging, or even complete interruption of the waterway.

In recent years (2014 to 2016) navigation had to be interrupted in the Paraná-Tietê waterway in Brazil due to extraordinarily low flows in those rivers. The interruption lasted 20 months and was related to the drought that gave rise to a water crisis in the São Paulo region in 2014 and 2015 (Nobre et al., 2016). A possible link between the São Paulo water crisis and climate change was investigated by Otto et al. (2015), who concluded that anthropogenic climate change was not the major influence on the drought of 2013/14. This conclusion is in accordance with our results which show that São Paulo and the Tietê River basin are within a region where there is no agreement between the GCMs on the direction of changes of both rainfall and river discharge.

Impacts on floods and droughts

A potential impact of climate change that is generally accepted as plausible is an increase in the magnitude and frequency of extreme events like floods and droughts. More intense rainfall events are expected, partly due to the increase in moisture carrying capacity of warmer air. It is not clear, however, how the intensification of rainfall events varies across temporal and spatial scales. Due to the methods used to downscale GCM projections to the basin scale, the assessment of climate change impacts presented here is inadequate to analyze the potential for extreme events, such as floods and droughts.

Adam et al. (2015) investigated potential changes in floods and droughts in the Paraná River Basin upstream of Itaipu using climate projections of a GCM downscaled using the Eta regional model developed by Chou et al. (2012). These authors concluded that natural climate variability can be as important as the influence of climate change in driving hydrological extremes. They investigated both high and low flow scenarios and found no clear pattern of increasing or decreasing magnitude or frequency in both floods and droughts. Camilloni et al. (2013), on the other hand, projected an increase in the frequency and duration of floods in the Paraná River at Corrientes. The contradiction between those previous assessments about floods may be related to the uncertainty associated with the GCMs used as sources for future climate scenarios.

Conclusions

The Paraná River Basin plays a key role in the economic development of South America, as the basin is home to around 90 million people and accounts for a large proportion of the energy and food production of Brazil, Paraguay and Argentina. While it has already been subject to considerable variability in precipitation and streamflow over the past decades, with relatively long periods above or below mean values, climate change may bring significant impacts on society and economic sectors by the end of this century. We evaluated long-term (2081–2100) climate change impacts on water resources of the Paraná River Basin using 25 Global Climate Models (GCMs) and scenarios from the IPCC AR5 report. The outputs of GCMs were used to force a continental-scale hydrologic-hydrodynamic model developed for the region of South America. According to GCM projections, precipitation will increase in the east and south of the basin by up to 10% and decrease in the north and northwest parts of the basin by up to 5%.

However, there is still a considerable disagreement between different GCMs in the magnitude of the changes in precipitation and even in the direction of change (positive or negative) in some regions.

Results of climate change impacts on hydrology obtained by the continental-scale hydrologic-hydrodynamic model show that there is general agreement that rivers that flow from north to south within the Paraná headwaters, that lie mostly within Brazil, will face discharge reductions from 5–30%. In particular, the Paraguay River may experience a larger decrease during the low flow period. There is no agreement, however, about the future changes in the discharge of the main Paraná River itself, except for a projected decline of about 5–10% in discharge in the Upper Paraná reach upstream of its confluence with the Paranapanema River in Brazil.

As a result of increased potential evapotranspiration, and reduced precipitation, the northern part of the Paraná River Basin is projected to become drier, as expressed by changes up to 0.5 in the Aridity Index (the ratio between potential evapotranspiration and precipitation), suggesting that activities such as agriculture and livestock production may be severely affected. Hydropower production is projected to face a reduction in the north, especially in the Paranaíba River, one of the tributaries of the Paraná. Along the main course of the Paraná to the south, a region that is shared by Paraguay, Argentina and Brazil where important hydropower plants like Itaipu and Yaciretá are located, there is no agreement among model projections about whether streamflow will increase or decrease. Navigation may be negatively impacted in the Paraguay River over its whole length, where projections show a future reduction in river discharge, and navigation may also be affected in the Upper Paraná within Brazil. On the other hand, navigation will probably not be affected by climate change in the lower reaches of the Paraná within Argentina, where there is no agreement among model scenarios about changes in river discharge.

Acknowledgements

We acknowledge the contributions of the following colleagues who participated in the efforts of developing the hydrological model of South America: Fernando Fan, Ayan Fleischmann, Aline Meyer (all from the Institute of Hydraulic Research – IPH/UFRGS). We also acknowledge efforts in previous assessment of climate change impact in the region by Paulo Rógenes Monteiro Pontes (Instituto Tecnológico Vale), Juan Martin Bravo (IPH/UFRGS) and Katiúcia Adam (FEEVALE University). Finally, we are thankful for the revisions by Chris Metcalfe, Vladimir Smakhtin and two anonymous reviewers.

References

Adam, K.N., Fan, F.M., Pontes, P.R., Bravo, J.M., Collischonn, W. 2015. Mudanças climáticas e vazões extremas na Bacia do Rio Paraná. *Revista Brasileira de Recursos Hídricos* 20(4):999–1007.

Amorim, P.B., Chaffe, P.B. 2019. Towards a comprehensive characterization of evidence in synthesis assessments: The climate change impacts on the Brazilian water resources. *Climatic Change* 155(1):37–57.

Antico, A., Kröhling, D.M. 2011. Solar motion and discharge of Paraná River, *South America: Evidence for a link. Geophysical Research Letters* 38(19):L19401.

Assad, E.D., Ribeiro, R.R.R., Nakai, A.M. 2019. Assessments and How an Increase in Temperature may Have an Impact on Agriculture in Brazil and Mapping of the Current and Future Situation. In: Nobre, C.A., Marengo, J.A., Soares, W.R. (Eds.) *Climate Change Risks in Brazil*, Springer, New York, NY, USA, pp. 87–98.

Barros, V., Clarke, R.T., Dias, P.L.S. 2006. *Climate change in the La Plata Basin*. Centro de Investigaciones del Mar y la Atmosfera, Buenos Aires, Argentina, 200 p.

Barros, V.R., Boninsegna, J.A., Camilloni, I.A., Chidiak, M., Magrín, G.O., Rusticucci, M. 2015. Climate change in Argentina: Trends. *projections, impacts and adaptation. Wiley Interdisciplinary Reviews: Climate Change* 6(2):151–169.

Beck, H.E., Van Dijk, A.I., Levizzani, V., Schellekens, J., Gonzalez Miralles, D., Martens, B., De Roo, A. 2017. MSWEP: 3-hourly 0.25 global gridded precipitation (1979-2015) by merging gauge, satellite, and reanalysis data. *Hydrology and Earth System Sciences* 21(1):589–615.

Berbery, E.H., Barros, V.R. 2002. The hydrologic cycle of the La Plata basin in South America. *Journal of Hydrometeorology* 3(6):630–645.

Boulanger, J.-P., Leloup, J., Penalba, O., Rusticucci, M., Lafon, F., Vargas, W. 2005. Observed precipitation in the Paraná-Plata hydrological basin: Long-term trends, *extreme conditions and ENSO teleconnections. Climate Dynamics* 24(4):393–413.

Bravo, J.M., Allasia, D., Paz, A.R., Collischonn, W., Tucci, C.E.M. 2011. Coupled hydrologic-hydraulic modeling of the Upper Paraguay River basin. *Journal of Hydrologic Engineering* 17(5):635–646.

Bravo, J.M., Collischonn, W., Paz, A.R., Allasia, D., Domecq, F. 2014. Impact of projected climate change on hydrologic regime of the Upper Paraguay River basin. *Climatic Change* 127:27–41.

Camilloni, I.A., Saurral, R.I., Montroull, N.B. 2013. Hydrological projections of fluvial floods in the Uruguay and Paraná basins under different climate change scenarios. *International Journal of River Basin Management* 11(4):389–399.

Chevallier, P., Pouyaud, B., Suarez, W., Condom, T. 2011. Climate change threats to environment in the tropical Andes: Glaciers and water resources. *Regional Environmental Change* 11:179–187.

Chou, S.C., Marengo, J.A., Lyra, A., Sueiro, G., Pesquero, J., Alves, L.M., Kay, G., Betts, R., Chagas, D., Gomes, J.L., Bustamante, J., Tavares, P. 2012. Downscaling of South America present climate driven by 4-member HadCM3 runs. *Climate Dynamics* 38(3–4):635–653.

CIESIN. 2018. *Gridded Population of the World, Version 4 (GPWv4): Population Count Adjusted to Match 2015 Revision of UN WPP Country Totals, Revision 11.* Palisades, NY: NASA Socioeconomic Data and Applications Center (SEDAC). Accessed July 2, 2019. doi:10.7927/H4PN93PB

Collischonn, W., Allasia, D.G., Silva, B.C., Tucci, C.E.M. 2007. The MGB-IPH model for large-scale rainfall-runoff modeling. *Hydrological Sciences Journal* 52:878–895.

Collischonn, W., Bravo, J.M., Pontes, P.R.M. 2016. Paraná River Basin. In: Singh, V.P., (Org) *Handbook of Applied Hydrology*, 2nd ed., McGraw-Hill Education, New York City, NY, USA, vol. 1, p. 1440.

Collischonn, W., Tucci, C.E.M., Clarke, R.T. 2001. Further evidence of changes in the hydrological regime of the River Paraguay: Part of a wider phenomenon of climate change? *Journal of Hydrology* 245(1–4):218–238.

Doyle, M.E., Barros, V.R. 2011. Attribution of the river flow growth in the Plata Basin. *International Journal of Climatology* 31(15):2234–2248.

Flato, G., Marotzke, J., Abiodun, B., Braconnot, P., Chou, S.C., Collins, W., Cox, P., Driouech, F., Emori, S., Eyring, V., Forest, C., Glecker, E., Guilyardi, E., Jakob, C., Kattsov, V., Reason, C., Rummukainen, M. 2013. Evaluation of climate models. In: *Climate Change 2013: The Physical Science Basis. Contribution of Working Group I to Fifth Assessment Report of the Intergovernmental Panel on Climate Change*, Cambridge University Press, Cambridge,UK, pp. 741–866.

Fleischmann, A.S., Siqueira, V.A., Paris, A., Collischonn, W., Paiva, R.C.D., Pontes, P.R.M., Crétaux, J.F., Bergé-Nguyen, M., Biancamaria, S., Gosset, M., Calmant, S., Tanimoune, B. 2018. Modelling hydrologic and hydrodynamic processes in basins with large semi-arid wetlands. *Journal of Hydrology* 561:943–959.

García, N.O., Vargas, W.M. 1996. The spatial variability of runoff and precipitation in the Rio de la Plata basin. *Hydrological Science Journal* 41:279–299.

García, N.O., Vargas, W.M. 1998. The temporal climatic variability of runoff and precipitation in the Rio de la Plata basin. *Climatic Change Journal* 38:359–379.

Genta, J.L., Perez-Iribarren, G., Mechoso, C.R. 1998. A recent increasing trend in the streamflow of rivers in southeastern South America. *Journal of Climate* 11(11):2858–2862.

Guerrero, M., Re, M., Kazimierski, L.D., Menéndez, Á.N., Ugarelli, R. 2013. Effect of climate change on navigation channel dredging of the Parana River. *International Journal of River Basin Management* 11(4):439–448.

Gulizia, C., Camilloni, I. 2015. Comparative analysis of the ability of a set of CMIP3 and CMIP5 global climate models to represent precipitation in South America. *International Journal of Climatology* 35(4):583–595.

Krepper, C.M., García, N.O., Jones, P.D. 2008. Low-frequency response of the upper Paraná basin. *International Journal of Climatology* 28:351–360.

Krepper, C.M., Venturini, V. 2009. Assessing interannual water balance of La Plata river basin. *Atmósfera* 22(4):387–398.

Lee, E., Livino, A., Han, S.C., Zhang, K., Briscoe, J., Kelman, J., Moorcroft, P. 2018. Land cover change explains the increasing discharge of the Paraná River. *Regional Environmental Change* 18(6):1871–1881.

Lima, J.W.M., Collischonn, W., Marengo, J.A. 2014. *Efeitos das mudanças climáticas na geração de energia elétrica, Report of AES Tiete S.A., São Paulo, Brazil, 360 p.*

Marengo, J.A., Bernasconi, M. 2015. Regional differences in aridity/drought conditions over Northeast Brazil: Present state and future projections. *Climatic Change* 129(1–2):103–115.

Marengo, J.A., Jones, R., Alves, L.M., Valverde, M.C. 2009. Future change of temperature and precipitation extremes in South America as derived from the PRECIS regional climate modeling system. *International Journal of Climatology* 29(15):2241–2255.

Melack, J., Coe, M. 2012. Climate change and the floodplain lakes of the Amazon basin. In: Goldman, C.R., Kumagai, M., Robarts, R.D. (Eds.) *Climate Change and Global Warming of Inland Waters: Impacts and Mitigation for Ecosystems and Societies*, John Wiley & Sons, Chichester, UK, pp. 295–310.

Montroull, N.B., Saurral, R.I., Camilloni, I.A. 2018. Hydrological impacts in La Plata basin under 1.5, 2 and 3°C global warming above the pre-industrial level. *International Journal of Climatology* 38(8):3355–3368.

Nobre, C.A., Marengo, J.A., Seluchi, M.E., Cuartas, L.A., Alves, L.M. 2016. Some characteristics and impacts of the drought and water crisis in Southeastern Brazil during 2014 and 2015. *Journal of Water Resource and Protection* 8(02):252.

Nóbrega, M.T., Collischonn, W., Tucci, C.E.M., Paz, A.R. 2011. Uncertainty in climate change impacts on water resources in the Rio Grande Basin, Brazil. *Hydrology and Earth System Sciences* 15:585–595.

Otto, F.E., Haustein, K., Uhe, P., Coelho, C.A., Aravequia, J.A., Almeida, W., Haarsma, R. 2015. Factors other than climate change, main drivers of 2014/15 water shortage in Southeast Brazil. *Bulletin of the American Meteorological Society* 96(12):S35–S40.

Paiva, R.C.D., Buarque, D.C., Collischonn, W., Bonnet, M.P., Frappart, F., Calmant, S., Bulhoes Mendes, C.A. 2013. Large-scale hydrologic and hydrodynamic modeling of the Amazon River basin. *Water Resources Research* 49(3):1226–1243.

Paiva, R.C.D., Collischonn, W., Tucci, C.E.M. 2011. Large scale hydrologic and hydrodynamic modeling using limited data and a GIS based approach. *Journal of Hydrology* 406:170–181.

Pontes, P.R.M., Fan, F.M., Fleischmann, A.S., de Paiva, R.C.D., Buarque, D.C., Siqueira, V.A., Jardim, P.F., Sorribas, M.V., Collischonn, W. 2017. MGB-IPH model for hydrological and hydraulic simulation of large floodplain river systems coupled with open source GIS. *Environmental Modelling and Software* 941:20.

Popescu, I., Brandimarte, L., Peviani, M. 2014. Effects of climate change over energy production in La Plata Basin. *International Journal of River Basin Management* 12(4):319–327.

Queiroz, A.R., Faria, V.A., Lima, L.M., Lima, J.W. 2019. Hydropower revenues under the threat of climate change in Brazil. *Renewable Energy* 133:873–882.

Pereira, A.F., de Lucena, A.F.P., Szklo, A.S., Schaeffer, R., de Souza, R.R., Borba, B.S.M.C., da Costa, I.V.L., Pereira Jr., A.O., da Cunha, S.H.F. 2009. The vulnerability of renewable energy to climate change in Brazil. *Energy Policy* 37:879–889.

Reyer, C.P.O., Adams, S., Albrecht, T., Baarsch, F., Boit, A., Trujillo, N.C., Cartsburg, M., Coumou, D., Eden, A., Fernandes, E., Langerwisch, F., Marcus, R., Mengel, M., Mira-Salama, D., Perette, M., Pereznieto, P., Rammig, A., Reinhardt, J., Robinson, A., Rocha, M., Sakschewski, B., Schaeffer, M., Schleussner, C.-F., Serdeczny, O., Thonicke, K. 2017. Climate change impacts in Latin America and the Caribbean and their implications for development. *Regional Environmental Change* 17(6):1601–1621.

Silveira, C.D.S., Vasconcelos Junior, F.D.C., Souza Filho, F.D.A.D., Guimarães, S.O., Marcos Junior, A.D., Dos Reis, G.N.L., Porto, V.C. 2019. Performance evaluation of AR5-CMIP5 models for the representation of seasonal and multi-annual variability of precipitation in Brazilian hydropower sector basins under RCP8. 5 scenarios. *Hydrological Sciences Journal*, 64:1279–1296.

Siqueira, V.A., Paiva, R.C.D.D., Fleischmann, A.S., Fan, F.M., Ruhoff, A.L., Pontes, P.R.M., Paris, A., Calmant, S., Collischonn, W. 2018. Toward continental hydrologic–hydrodynamic modeling in South America. *Hydrology and Earth System Sciences* 22(9):4815–4842.

Sordo-Ward, A., Bejarano, M., Iglesias, A., Asenjo, V., Garrote, L. 2017. Analysis of current and future SPEI droughts in the La Plata basin based on results from the regional Eta climate model. *Water* 9(11):857.

Stocker, T.F., Qin, D., Plattner, G.-K., Tignor, M., Allen, S.K., Boschung, J., Nauels, A., Xia, Y., Bex, V., Midgley, P.M. 2013. *Climate change 2013: The physical science basis. Contribution of Working Group I to the Fifth Assessment Report of the Intergovernmental Panel on Climate Change.* Cambridge: Cambridge University Press, 1535 p.

Thomson, A.M., Calvin, K.V., Smith, S.J., Kyle, G.P., Volke, A., Patel, P. 2011. RCP4.5: A pathway for stabilization of radiative forcing by 2100. *Climatic Change* 109(1):77–94.

Tucci, C.E.M., Clarke, R.T. 1998. Environmental issues of La Plata Basin. *Water Resources Development* 14:157–173.

Tundisi, J.G., Matsumura-Tundisi, T. in press. Water infrastructure within the basin. In: Metcalfe, C.D., Collins, P., Tundisi, J.G., Menone, M. (Eds.) *The Paraná River Basin: Managing Water Resources to Sustain Ecosystem Services*, Smakhtin, V. (Series Editor), Earthscan Series on Major River Basins of the World, Routledge.

Vuille, M., Carey, M., Huggel, C., Buytaert, W., Rabatel, A., Jacobsen, D., Soruco, A., Villacis, M., Yarleque, C., Timm, O.E., Condom, T., Salzmann, N., Sicart, J.-E. 2018. Rapid decline of snow and ice in the tropical Andes – Impacts, uncertainties and challenges ahead. *Earth-Science Reviews* 176:195–213.

Yang, Y., Zhang, S., McVicar, T.R., Beck, H.E., Zhang, Y., Liu, B. 2018. Disconnection between trends of atmospheric drying and continental runoff. *Water Resources Research* 54(7):4700–4713.

Zullo, J., Pinto, H.S., Assad, E.D., de Ávila, A.M.H. 2011. Potential for growing Arabica coffee in the extreme south of Brazil in a warmer world. *Climatic Change* 109(3–4):535–548.

7

Water quality and impacts of development

Mirta L. Menone and Franco Teixeira de Mello

Key messages

- Although the Upper Paraná contributes >50% of the water budget of the river, the largest proportion of transported suspended sediment and organic material is carried into the Middle Paraná from the Paraguay River. Increased sediment load constitutes a stressor and is caused by changes in land-use and land-based pollution. Surface runoff from urban centers constitutes a source of nutrients all along the river, while fish farming and cattle ranching have been contributing nutrients in the Upper and Middle stretches, respectively.

- Anthropogenic activities are responsible for impaired water quality in the Paraná River Basin. The water quality of the Upper Paraná basin is affected by transport of pesticides from agricultural activities and by discharges from industry. However, the most influential anthropogenic impacts are from the approximately 140 hydroelectric dams and reservoirs. Inputs of nutrients into these reservoirs are causing eutrophication, which is resulting in harmful blooms of cyanobacteria. The Middle and Lower stretches of the Paraná River Basin are affected by discharges of partially treated domestic wastewater and industrial pollution, as well as agricultural activities.
- High concentrations of metals and metalloids such as Cd, Pb, Cu, F, Ni, Cr and As have been detected in some reaches of the river. While for some of these elements, such as As, natural geochemical processes could be the main source, others originate from the effluents of metallurgical and tannery industries and wastewater discharges from cities. The concentrations of Cd, Pb, Cr, Fe and Mn in sediments often exceed regulatory limits for the protection of aquatic life.
- Current use insecticides, the persistent organochlorine insecticide endosulfan and the herbicide glyphosate are ubiquitous in the Paraná River Basin. In the Upper Paraná stretch the main pesticides reported are the insecticides malathion, lambda cyalothrina, endosulfan and the herbicide metolachlor. In the Middle and Lower Paraná stretches, the insecticides cypermethrin and chlorpyrifos and the herbicide glyphosate predominate in surface waters. A recent study reported concentrations of pesticides in water in the basin ranging from 0.004–6.62 $\mu g\ L^{-1}$, with concentrations of some compounds often over the recommended guidelines for the protection of aquatic biota.

Introduction

According to UNSD/UNEP (2004) the 'categories' needed to monitor basic water quality include annual average flow (for rivers), carbonaceous biochemical oxygen demand ($cBOD_5$), chemical oxygen demand (COD), dissolved oxygen (DO), total dissolved solids (TDS), total phosphorus (TP), total nitrogen (TN) and levels of fecal coliforms. In order to assess progress in meeting the Sustainable Development Goal for 'good quality' of surface waters (i.e. SDG 6.3.2), UN-Water (2017) recommends monitoring the six 'core' parameters of pH, conductivity, DO, total oxidized nitrogen (TON) and orthophosphate (OP), and also recommends measuring several other 'progressive' parameters, such as turbidity, hardness, alkalinity, suspended solids, BOD, COD, nitrate, total phosphorus, major anions and cations, metals, hydrocarbons, pesticides, *E. coli* and fecal coliform bacteria. In the case of the Paraná River Basin, records of $cBOD_5$ levels show that this parameter might not be a relevant water quality indicator, as the levels of biological oxygen demand are consistently near the method detection limit (Natale, 2005). The levels of dissolved toxic metals and organic pollutants are also frequently under the method detection limits, although they are relevant water quality indicators for monitoring point-sources of contamination. For instance, metal contamination and silt originating from mining operations in the headwaters of the Bolivian Andes may be impacting water quality further downstream in the Pilcomayo River sub-basin (Natale, 2005).

As described in Chapter 1 in this book, the Paraná basin is subdivided into several regions and sub-basins, including the Paraguay River system, the Upper Paraná, the Middle Paraná and the Lower Paraná (including the Delta region). The water quality varies widely across these regions, depending on the hydrologic, geological and anthropogenic conditions. The Upper Paraná is the most significant contributor to the water budget for the system, but as described in Chapter 4 in this book, the extensive network of reservoirs within this system significantly reduces the levels of suspended sediments. Therefore, the largest proportion of transported suspended sediment is carried into the Middle Paraná from the Bermejo River drainage, which represents less than 2% of the total area of the basin. Most of the organic material that enters the Middle Paraná is provided by the Paraguay River, which delivers a greater load of particulate organic carbon than the entire Upper Paraná system (Depetris and Pasquini, 2007). In the Upper Paraná, many basic water quality parameters (e.g. pH, conductivity and alkalinity) remain relatively constant, independent of annual oscillations in hydrometric conditions. Total nitrogen and phosphorus concentrations tend to increase with rising hydrometric levels in the Paraná River, which is the opposite of what is observed in the floodplain environments where nutrient levels decline during flood events (Agostinho et al., 2009). Chemical contaminants also contribute to deterioration in water quality in the river, including runoff of pesticides from the agricultural activities and inputs of metals and aromatic hydrocarbons from urban and industrial activities (Weneck Lima et al., 2005).

The Upper Paraná basin in Brazil, particularly in the region between the city of Três Lagoas in the State of Mato Grosso do Sul and the city of Guairá in the State of Paraná, has the greatest density of populations and industry. This has contributed to gross pollution of the river systems, as illustrated by the many reports of pollutants of industrial and domestic origin entering the Tietê River (Figure 7.1). In addition, there is intensive agriculture in this region, with extensive use of agrochemicals, especially along the tributaries. In São Paulo State in Brazil, there is

Figure 7.1 Foam entering the Tietê River as a result of the discharge of surfactants and detergents into the river from industry and domestic wastewater.

Photograph obtained through an agreement with dropbox.com.

a high level of water consumption for urban, agriculture and industrial purposes, and this water demand receives little treatment before returning to the river. The intense use of pesticides and fertilizers for agriculture in the Upper Paraná basin has contributed to poor water quality in the river as well as its tributaries (Agostinho et al., 2003).

In the Upper Paraná, hydroelectric dams are present in all major tributaries, including the Grande, Paranaíba, Tietê and Paranapanema Rivers. Therefore, the river is regulated by both large and small reservoirs that affect almost 70% of the area of the basin (Agostinho et al., 2000, 2003). These dams and reservoirs are the most influential anthropogenic source of changes in water quality. Presently, there are approximately 140 reservoirs in total in operation with a dam height >10 m, of which 14 dams are higher than 150 m and are classified as 'high dams' (Isupova and Mikhailov, 2018). These 14 high dams are mentioned among the 46 large dams in the Upper Paraná that are described in Chapter 4 in this book. The Itaipú dam is the second largest hydropower facility in the world, surpassed only by the Three Gorges Dam in China.

The control of water levels by the dams located upstream of stretches of floodplain influences nutrient cycling and the structure and composition of biological assemblages (Agostinho et al., 2009). Damming of stretches of the river has increased levels of nutrients and reduced dissolved oxygen. This has caused changes in the phytoplankton community, and in particular, has resulted in blooms of cyanobacteria that cause a significant reduction in water quality. Cyanobacteria blooms have been observed in several parts of the Upper Paraná, and cyanotoxins (i.e. microcystins) have been detected in the Itaipú Reservoir (Agostinho et al., 2009). Other cyanotoxins (i.e. saxitoxin) have been accumulated in drinking water and in shellfish to levels that are a threat to public health through paralytic shellfish poisoning (Clemente et al., 2010; Carmichael and Boyer, 2016; Le Moal et al., 2019). In the stretches of the Upper Paraná that have not been dammed, water quality is still influenced by changing water levels induced by upstream dams, as well as inputs of nutrients, sediments and agrochemicals from cattle-ranching, rice crop irrigation, extraction of ginseng and sand and aggregate mining (Agostinho et al., 2003).

In the upper basin of the Paraguay River, the Pantanal wetland is a huge lowland of approximately 150,000 km^2. Water within the Pantanal is eventually released into the basin. However, the long time-span for water retention within the Pantanal and other wetlands in the basin can affect water quality in the main stem of the river at different periods in the hydrological cycle. One of the well-known phenomena related to water quality in the basin is regionally known as '*dequada*' or '*decoada*', which is associated with the return of floodplain water to the main channels, which causes reductions in water transparency, higher pH and conductivity and reduced levels of dissolved oxygen (Bergier and Assine, 2016).

The Middle Paraná starts downstream of the confluence of the Upper Paraná River with the Paraguay River at Corrientes. The water quality in the Middle Paraná region is predominantly affected by discharges of domestic wastewater, industrial pollution and agricultural activities (Blettler et al., 2018). According to Bonetto (1986), the water quality of the Middle Paraná, near the cities of Santa Fe and Paraná, is quite different from that of the Upper Paraná. The average conductivity, 90 μS cm^{-1}, is twice that of the Upper Paraná. Bicarbonate dominates the anions (average > 35 mg L^{-1}), while levels of chloride (average 10 mg L^{-1}) are higher and sulphate concentrations (average 5 mg L^{-1}) are lower in the Middle Paraná relative to the Upper Paraná.

The Lower Paraná River downstream of the city of Santa Fe is surrounded by densely populated areas, including the major urban centres of Rosario, Zárate and Campana, and in these cities, there are tanneries, petrochemical plants and pharmaceutical, metallurgic and other industries. Industrial effluents, as well as domestic wastes, are flushed into the Lower Paraná

River with little or no treatment (Cataldo et al., 2001). Therefore, industrial activities and discharges of untreated or partially treated domestic wastewater are the main contributors to the deterioration of water quality in the Lower Paraná. A recent study by Sica et al. (2016) also described the impact of land use changes on the Lower Paraná, mainly due to the conversion of natural areas to grazing lands. One third of the freshwater marshes of the Lower Delta (i.e. 163,000 ha) have been replaced by pastures (70%) and forested areas (18%) over a time-span of only 14 years (Sica et al., 2016). These rapid and widespread losses of freshwater marshes have potentially negative consequences for ecosystem services, water quality and biodiversity.

All sections of the Paraná River Basin receive inputs of agrochemicals from runoff, which is a ubiquitous source of contamination over the entire length of the river system. In particular, the Middle and Lower stretches of the Paraná tributaries contribute significant loads of pesticides, and, in spite of high dilutions within the river, these pesticides are detectable in the main watercourse. The high level of the organochlorine insecticide, endosulfan in water and sediments in the basin is an example of contamination by a pesticide that has been banned but is still making its way into surface waters at levels that may be a hazard to aquatic organisms and to human health (Etchegoyen et al., 2017).

This chapter summarizes the data currently available in the peer-reviewed literature and to some extent in the 'grey' literature (e.g. graduate theses, technical reports, conference proceedings) on water quality parameters within the Paraná River Basin. Because of the differences in hydrological characteristics, geology and land-use patterns within the various regions of the basin, some of these water quality data are concentrated within specific geographic regions. Brazil, and to a lesser extent, Argentina are monitoring water quality at various sites and are reporting these data to the global GEMStat database. However, we could not access these data, as they are proprietary to the country of origin and to the United Nations agency, GEMS/Water (UN-Water, 2017).

Suspended sediments

The levels of suspended sediments in water are an important water quality parameter because of the interactions between sediments and the major anions, such as carbonate and bicarbonate that are key regulators of water alkalinity (Abarca, 2007). Some metal ions, as well as organic matter contribute to the formation of aggregates of suspended particles and are transported with sediments in fluvial systems (Suárez, 2012). In addition, high sediment loads may also be a significant stressor for aquatic organisms (Larson et al., 2019). Rivers located in tropical and sub-tropical ecozones, such as the Paraná River, play key roles in the storage of sediment at a continental scale and also contribute at a global scale to the transport of sediment and nutrients into the oceans. The Paraná River system is important for transporting sediments from the South American continent to the Atlantic Ocean through the Rio de la Plata estuary. However, these aquatic systems are at high risk of increased sediment loads because of changes in land-use, deforestation and land-based pollution, as well as construction of dams (Latrubesse and Abad, 2018). Climate conditions also influence these river systems. For instance, from the second half of the 20th century to the present, precipitation in the Paraná River increased by 10% over the entire basin and by 30% for some areas of the basin, causing a rise in the levels of soil saturation and increased runoff (Isupova and Mikhailov, 2018). Currently, the hydro-sedimentological dynamics of the Paraná River are primarily under the control of a series of dams, which has transformed the main channel of the Paraná River into a sequence of reservoirs (Pinto Martins et al., 2009).

Drago and Amsler (1998) described the bed sediment composition of the Paraná River Basin. The main channel has a sandy mobile bed composed of quartzose sands plus the fine fractions of sediment supplied primarily by the Bermejo River that discharges into the lower Paraguay River. In the Lower Paraná, a large delta of 345 km length has developed, with a multilobate shape that reflects the strong transport and deposition dynamics of the Paraná River. In the Delta, the Paraná River divides into two main branches: the Paraná Guazú and the Paraná de las Palmas. The Paraná Guazú branch captures most of the water discharge and the sediment load, presenting the characteristics of a true alluvial river with a predominantly sandy bed up to the Río de la Plata estuary. On the other hand, the Paraná de las Palmas branch has a bed sediment with a high clay and silt content over nearly all of its length. The characteristics of bed sediments and suspended sediments in the basin are described in more detail in Chapter 3.

Estimates made by Weneck Lima et al. (2005) indicate that the suspended sediment loads generated only within the Brazilian portion of the Upper Paraná basin amount to about 10.5×10^6 t yr^{-1}. These relatively small loads indicate that a large proportion of the sediments are trapped by reservoirs. However, Pinto Martins et al. (2009) estimated that the average bedload transport in the Upper Paraná River, between Porto Primavera and the Itaipú dam was approximately 1.2×10^6 t yr^{-1}, one order of magnitude lower than the values estimated by Weneck Lima et al. (2005).

In the Middle Paraná, most of the sediment load comes from the Paraguay River, which supplies approximately 92% of the total load (Maine et al., 2004). The remaining 8% of the sediments that enter the Middle Paraná come from the Upper Paraná. The Bermejo River and Pilcomayo River sub-basins are primarily responsible for the high sediment loads observed in the Paraguay River, since the Pilcomayo River carries about 9.8×10^7 t yr^{-1} of sediments (Facetti-Masulli and Klump, 2010) and the Bermejo River delivers more than 1×10^8 t yr^{-1} of sediments (Orfeo and Iriondo, 2012). The mean concentration of total dissolved solids of 45 mg L^{-1} is within the range of some other large rivers in the world, such as the Amazon and the Congo (Campodonico et al., 2015). Chapter 3 in this book provides more details about the reduction in the mean concentrations of suspended sediments from the pre-dam period to the values observed today.

According to Milliman and Farnsworth (2011), as reported in Campodonico et al. (2015) the Paraná River transports 6.2×10^4 t yr^{-1} of dissolved solids and 9×10^7 t yr^{-1} of total suspended solids to the Atlantic Ocean. However, according to Iriondo (2004), the Lower Paraná annual discharge of suspended solids into the Atlantic Ocean is about 200×10^6 t yr^{-1}. As can be seen, the estimates of sediment load in the Paraná River Basin are not always consistent with each other. While a total sediment load of 90×10^6 t yr^{-1} has been reported for the entire river, a load of 120×10^6 t yr^{-1} has been estimated only for the Middle Paraná and 100×10^6 t yr^{-1} for only the Bermejo River (Pinto Martins et al., 2009; Campodonico et al., 2015). Although a portion of these inconsistencies can arise from different periods of study, these types of discrepancies are also observed in the literature concerning other hydrological parameters, such as the variations in water flow (Isupova and Mikhailov, 2018).

Major ions and nutrients

In the waters of the Paraná River, the general order of abundance of major anions is $HCO_3^- > Cl^- > SO_4^{2-}$, and the relative abundance of major cations is $[Na^+ + K^+] > Ca^{2+} > Mg^{2+}$, which is within the normal range expected in this type of fluvial system (Depetris and Pasquini, 2007). In the Upper Paraná basin between Três Lagoas city in the State of Mato Grosso do Sul and Guairá city in the State of Paraná, the concentrations of total phosphorus average 23 μg L^{-1} and 64 μg L^{-1} in the river and in floodplain lagoons, respectively (Agostinho et al., 2000). As benchmarks for assessing the threat of eutrophication, the US EPA has established guidelines for total phosphorus of 100 μg L^{-1} for streams and rivers and 25 μg L^{-1} for lakes and reservoirs. For the Upper Parana, Barros et al. (2017) observed that in Xambrê and Pinhalzinho rivers that receive surface runoff from Umuarama city, the urban sewage and industrial wastewater produce adverse effects in fish, such as liver and gill histopathological alterations in *Astyanax spp*. These rivers that are also utilized for public water supply have very poor water quality. In particular, the Pinhalzinho River has high levels of total nitrogen (7.6±0.8 mg L^{-1}) and ammonia (3.88±0.18 mg L^{-1}), high turbidity (11.5±0.3 NTU) and very low oxygen levels (0.7±0.1 mg L^{-1}).

Trapping of phosphate in upstream reservoirs may explain the relatively low levels of phosphate in the Upper Paraná River. On the contrary, there are high concentrations of nitrate (25–232 μg L^{-1}) in the river, in comparison to the concentrations of this anion (<25 μg L^{-1}) in the floodplain lagoons (Agostinho et al., 2000). In a study of nutrient loads in the Passaúna reservoir in the Metropolitan Region of Curitiba in the Upper Paraná, concentrations of nutrients were extremely high, with levels of total phosphorus in water between approximately 100 to over 2,000 μg L^{-1} and the concentrations of total nitrogen in the range of 500–2,000 μg L^{-1} (Veiga and Dziedzic, 2010).

Several studies conducted in Brazil have described adverse effects on water quality from fish farming, including inputs of organic carbon and nutrients. In the Upper Paraná basin, there is evidence that the effluents from rural fish farms are causing a negative effect on water quality. In a study of three headwater streams situated in the Federal State of Minas Gerais (Brazil), discharges of fish farm effluents resulted in up to a 2-fold increase in stream particulate organic matter and up to a 1.8-fold increase in inorganic nutrients, resulting in effects on the ecosystem metabolism (Dos Santos Rosa et al., 2013). Farming of fish is also widespread in reservoirs in the Upper Paraná and inputs of nutrients from these practices are contributing to eutrophication. In reservoirs on the Paranapanema River in Brazil, fish farming was estimated to contribute a total phosphorus load of >194 kg ha^{-1} yr^{-1} to the system (Neto et al., 2017).

Some studies in this area have reported variations in water quality due to 'El Niño' climatic cycles. During a southern oscillation (i.e. ENSO) event in 1982–1983, there were lower levels of total dissolved solids (TDS) and oxygen supersaturation in the Middle Paraná with respect to average values (Depetris and Pasquini, 2007). Over almost the same period of time as this ENSO event, data generated for the Paraná within the framework of the SCOPE-UNEP International Carbon Project indicated a mean transport rate of total organic carbon (TOC) of 8–9 Tg y^{-1}, of which ~85% corresponded to dissolved organic carbon and ~15% to particulate organic carbon (Depetris and Kempe, 1993).

In the Middle Paraná, high concentrations of ammonium, nitrates and phosphates, as well as other pollutants have been found in surface waters near urban areas (Blettler et al., 2018). Maine et al. (2004) conducted a one-year study along a 400 km stretch of the Middle Paraná and in five floodplain lakes. Water quality data showed that the waters of the Paraná River were circumneutral in pH and well oxygenated (i.e. dissolved oxygen concentrations close to saturation), and the mean concentrations of nitrates and soluble reactive phosphorus (SRP) in water were ~27 and 210 μg L^{-1}, respectively. Particularly along the San Javier River sub-basin, high inorganic nitrogen concentrations indicated contributions from point-sources (e.g. urban sewage, discharges from dairy and chicken farms, etc.). In contrast, increased SRP concentrations corresponding with flood events indicated non-point-source contributions of runoff from agriculture (Maine et al., 2004). However, the authors of this study concluded that floodplain lakes in the Middle Paraná are a trap for nutrients and partially counteract upstream inputs of nutrients from urban centers.

Another reason for the increase in nutrient levels could be the introduction of cattle ranching in the floodplains. In Argentina, the expansion of soybean production forced the relocation of cattle to more marginal sites, such as wetlands. Consequently, Mesa et al. (2015) observed in the Middle Paraná area, a significant increase in SRP, total phosphorus and total nitrogen in water corresponding with the introduction of ranching, with concentrations of these nutrients reaching values between two and three times higher than those reported in other wetlands of the Paraná River system. In addition, cattle could also remobilize nutrients by re-suspending nutrient-bound sediments as the animals walk within wetlands. Sediment resuspension by cattle trampling could also explain the increase in the concentration of chromophoric dissolved organic matter (Mesa et al., 2015).

Bonetto et al. (1994) reported values for nutrients in the Lower Paraná River, including mean total phosphorus concentrations of 338 μg L^{-1}, mean SRP concentrations of 24 μg L^{-1} and mean concentrations of nitrate + nitrite of 195 μg L^{-1}, indicating risks of eutrophication. These researchers also measured these parameters in a floodplain lake in contact with the river and the surrounding marsh in the delta of the river. There was a large decrease in the concentrations of suspended matter, nitrate + nitrite and SPR in a gradient from the river to the lake and marsh. These researchers noted that samples collected from the Lower Paraná, in terms of total phosphorus, resembled those of the Upper Paraná River for samples collected in November 1990 and January 1991 (i.e. during flooding), and resembled those of the Bermejo River for samples collected throughout the rest of the year. However, in a study conducted 10 years later, Agostinho et al. (2000) detected concentrations of total phosphorus in the Upper Paraná that were one order of magnitude lower. Therefore, it is difficult to establish a range of nutrient concentrations throughout the Paraná River, because of temporal differences between the various studies and variability in the inputs from anthropogenic sources of pollution.

Finally, Villar et al. (1998) studied changes in water chemistry in the lower stretch of the Paraná River during an ENSO event in 1997–1998, and they found that the transport of total phosphorus and SRP into the river increased during the ENSO event by 42% and 105%, respectively. They also observed enrichment of organic carbon during the flood. During an ENSO event in 2009–2010, water quality monitoring in the Rio de la Plata estuary indicated that flows increased, Secchi depth decreased and nutrient concentrations increased (Brugnoli et al., 2019).

Metals and metalloids

Spikes in the concentrations of metals associated with particles suspended in water are usually correlated with flood events that resuspend bed sediments (Peraza-Castro et al., 2016). Areas in fluvial systems where water velocity is reduced favor the deposition of fine particles of sediment and associated metal contaminants. In the Amambaí River in the Upper Paraná River Basin in the Brazilian state of Mato Grosso do Sul (MS), high concentrations of Cd, Pb, Cr, Cu, F were found in sediment samples; values that are higher than the limits set by CONAMA (Viana et al., 2017). Most of these metals were also found in the fish species, *Astyanax lacustris*, and their presence was associated with mutagenic and genotoxic effects (Viana et al., 2017, 2018).

The Iguazú River sub-basin of the Upper Paraná provides the majority of the water supply (i.e. 80%) for the Metropolitan Region of Curitiba (de Andrade Brito et al., 2018).

After passing through this large urban region, the Iguazú River is considered the second most polluted urban river in Brazil because of the discharges of domestic sewage and industrial wastewater that contribute pollutants to the river, including toxic metals. According to monitoring data collected by the Brazilian National Environment Council, the concentrations of copper, manganese and iron in the river (Table 7.1) were elevated above the regulatory limits 0.009, 0.1 and 0.18 mg L^{-1}, respectively. The Barigui River, which is a tributary of the Iguazú River is also strongly impacted by urban development. While the concentrations of dissolved metals in the river water were low, the concentrations of lead, chromium, nickel and zinc in bottom sediments were high and reached ≥ 50% of the sediment quality guidelines for Brazil (Resolução Conselho Nacional do Meio Ambiente – CONAMA 344/04) (Furtado Gonçalves et al., 2016).

Table 7.1 Concentrations of metals, pesticides and organic contaminants in surface waters at selected sites in the Paraná River Basin.

Area of the Basin	*Type of Contaminant*	*Concentration (μg L^{-1})*	*Reference*
Upper Paraná River	Aromatic hydrocarbons	3.6 $\sum$ PAHs	de Andrade Brito et al. (2018)[1]
	Metals	11.8 Cu	
		528.56 Mn	
		145.8 Zn	
		138.86 Fe	
		16.00 Pb	
	Current use pesticides	< 0.3–0.5 glyphosate	Ronco et al. (2016)[2]
		< 0.3 AMPA (glyphosate´s metabolite)	
		0.01–0.17 malathion	Laabs et al. (2002)[3]
		0.004–0.014 metolachlor	
		0.007–0.019 alachlor	
		0.003–0.019 trifluralin	
		0.003–0.004 endosulfan- sulfate	
		0.008–0.01 atrazine	
	Organochlorine pesticides	0.009 lindane	Lenardón et al. (1984)[4]
	Organophosphate pesticides	0.009 benzene hexachloride	
		0.022 parathion	
Middle Paraná River	Metals	12.1 Cr	Gallo et al. (2006)[5]
		1306 Fe	
		384 Mn	
	Glyphosate	0.6 (A)	Ronco et al. (2016)[6]
	AMPA (glyphosate´s metabolite)	< 0.3 (D)	
		0.7 (E)	
		< 0.3 (A, D, E)	
Lower Paraná River	Aliphatic hydrocarbons	3.44 $\sum$ *n*C15, *n*C16, NorPristane, *n*C24, *n*C25	Cataldo et al. (2001)[7]
	Metals, dissolved	3.96 Cr	
		72.70 Zn	
		3.87 Cu	
	Metals, suspended particulates	303.00 Cr (dry weight)	
		55.90 Cu (dry weight)	
		115.00 Pb (dry weight)	

Current use pesticides	0.37 α-cypermethrin (floodwater) (B) 0.19 α-cypermethrin (run-off water) (B) 0.33 chlorpyrifos (floodwater) (B) 0.18 chlorpyrifos (run-off water) (B) 0.15 chlorpyrifos (run-off water) (H)	Jergentz et al. (2005)[8*]

[1] Iguazú River
[2] Pilcomayo River, Paraguay River (main watercourse and confluence with the Paraná River), Montelindo stream and Bermejo River, Negro River, Santa Lucía River, Paraná River (main watercourse downstream the cities of Resistencia and Corrientes), Corrientes River, Guayquiraró River, Feliciano River and Salado River (near Santa Fé and Santo Tomé cities
[3] Rivers of the northeastern Pantanal basin
[4] Near Santa Fe and Paraná cities
[5] Salado River
[6] A: San Lorenzo River, D: Saladillo River, Paraná River (main watercourse), Pavon stream, Ramallo stream, Arrecifes River, Río Areco, Paraná de las Palmas River; E: Luján River (tributary of the Paraná Delta-Río de la Plata system)
[7] Paraná de las Palmas
[8] Brown stream (B) and Horqueta stream (H)
* mean values.

Peluso et al. (2013) reported high concentrations of Cd, Pb, Hg and Zn in bottom sediments in the Middle Paraná. The Salado River, which discharges into the Middle Paraná system has headwaters located in the Salta and Catamarca states of Argentina, and it passes along a stretch of about 1,500 km until reaching the confluence with the Santa Fe River. Human activities like dairy production, agriculture and industry result in significant impairments to water quality as a result of inputs from both point and non-point sources. Metal contamination originates from the effluents of metallurgical and tannery industries located in Santa Fe, Rafaela and Esperanza cities, among other urban centers. The metals Cr, Fe and Mn were detected in samples of water collected from the Salado River (Table 7.1), and mass flows were calculated showing a load of heavy metals at levels of 0.7 to 2.2 kg/s of Fe, 20 to 60 g/s of Mn and 1 to 3 g/s of Cr. In this study, the total Cr average concentration varied from <10 to 30 $\mu g\ L^{-1}$, higher than the Argentinean Guideline for protection of aquatic biota of 2.5 $\mu g\ L^{-1}$ (Desarrollos de Niveles Guía Nacionales, 2003). All metals were found predominantly in particulate form, with very low values for dissolved fractions (Gallo et al., 2006).

Blettler et al. (2018) reported that bottom sediments collected from several sites downstream of urban sewage discharges into the Colastiné River, which is a major secondary channel of the Paraná River situated in Santa Fe province in Argentina, contained high concentrations of several metals and metalloids (i.e. Cu, Pb, Zn, Al, Ba, B, Co, Fe, K, Li, Mg, Mn, P, Se, Sr and V). In particular, Blettler et al. (2018) noted that As, Cr and Ni were present in sediments at much higher concentrations than the sediment quality guidelines listed by Burton (2002). It is evident that wastewater discharges from the city of Santa Fe increase the metal concentrations of Cr and Ni. However, high As concentrations were attributed to natural geochemical processes (Blettler et al., 2018). Within the Paraná River Basin, there are thick beds of marine and continental sedimentary rocks, as well as outcrops of metamorphic rocks along with volcanic rocks of the Quaternary period that could be a geologic source of metals (Pasquini and Depetris, 2012). On the other hand, anthropogenic sources also contribute to metal contamination in the Middle Paraná system from the large urban centers and factories within Argentina and Paraguay (Peluso et al., 2013).

Pesticides and other organic contaminants

The intensive agriculture throughout the Paraná River Basin and its tributaries contribute significant loads of pesticides that reach the main watercourse. A recent study by Etchegoyen et al. (2017) that included monitoring in the upper, middle and lower stretches of the Paraná River Basin showed that the ranges of total measured pesticide concentrations in water and sediments were, respectively, 0.004–6.62 μg L^{-1} and 0.16–221.3 μg kg^{-1} dry weight (dw). The insecticides endosulfan, cypermethrin and chlorpyrifos were ubiquitous compounds in both environmental compartments, and the concentrations of these compounds detected in water were often over the recommended guidelines for the protection of aquatic biota (Etchegoyen et al., 2017).

In the Upper Paraná, and particularly in the Mato Grosso do Sul state of Brazil, several areas are highly impacted by agrochemicals from sugarcane production. For example, 2.28 million hectares, or the equivalent of 47% of the land area of the Ivinhema River sub-basin is under sugarcane cultivation (Frata and Batolomeu de Faria, 2010). Also, burning of the sugarcane before harvest might contribute to the deposition of ash to water courses, which may be toxic to native fish species, favoring the spread of exotic species (Gonino et al., 2019). In a recent study in São Lourenço River in Brazil that discharges into the Pantanal region, and is a region with intensive agricultural activity, 41% of the 39 samples of surface water analyzed showed detectable residues of pesticides. The most prevalent compounds were the insecticides carbofuran, malathion, permethrin, imidacloprid and acetamiprid, the herbicides atrazine, trifluralin, diuron and metolachlor, and the fungicide carbendazim (Bertón, 2017). In the Cuiabá River, which is a major tributary of the Paraguay River sub-basin located downstream of the São Lourenço River, analysis of pesticides in surface water indicated that metolachlor (herbicide) and flutriafol (fungicide) were detected in less than 5% of samples collect from 13 points monitored in 2010/2011 (Batista de Figueiredo, 2012). Also, in surface water samples collected from the Cuiabá River, Possavatz (2013) detected the presence of endosulfan (0.36 μg L^{-1}), clomazone (2.8 μg L^{-1}) and malathion (0.54 μg L^{-1}) in four samples among 180 samples collected monthly from August 2011 to July 2012. In samples of bottom sediments, the insecticides with maximum concentrations were metolachlor (21.2 μg kg^{-1}), permethrin (44.5 μg kg^{-1}), lambda cyhalothrin (21.2 μg kg^{-1}), alpha-endosulfan (13.6 μg kg^{-1}), endosulfan sulphate (79.3 μg kg^{-1}), chlorpyrifos (8.5 μg kg^{-1}) and malathion (9.2 μg kg^{-1}) These pesticides exhibited variations in their concentrations over time, which highlights the importance of understanding the spatial and temporal dynamics in different environmental matrices.

In the northeastern part of the Pantanal basin, pesticides were detected in stream and river waters at concentrations below 0.1 μg L^{-1} and in sediments at concentrations below 10 μg kg^{-1} dw (Bergier and Assine, 2016). Pesticides with higher detection frequencies included herbicides from the triazine class (i.e. atrazine, simazine, ametryn) and the acetanilide class (i.e. alachlor, metolachlor) and trifluralin, as well as the organochlorine insecticide, β-endosulfan and its metabolite, endosulfan sulfate. In an earlier study from the Pantanal, some organophosphorus pesticides (e.g., monocrotophos, profenofos, triazophos) were detected close to the site of application but they were rarely detected outside the application area (Laabs et al., 2002).

In a recent study of small streams that flow through agricultural fields in two regions of soybean production, namely, the tributaries of the Pirapó River in Paraguay and tributaries of the San Francisco River in Brazil, the maximum concentrations of insecticides detected in sediments were 1.47 ng g^{-1} dw (i.e. 1.47 μg kg^{-1}) for chlorpyrifos, 0.85 ng g^{-1} dw (i.e. 0.85 μg kg^{-1}) for endosulfan, 4.94 ng g^{-1} dw (i.e. 4.94 μg kg^{-1}) for cypermethrin and 16.57 ng g^{-1} dw (i.e. 16.57 μg kg^{-1}) for lambda cyhalothrin (Hunt et al., 2016). Banned organochlorines were most frequently detected in Brazil (Hunt et al., 2016). For example, endosulfan has been banned since 2013 in the region, but it is still detected in these watersheds.

Other organic contaminants like the anti-inflammatory drugs diclofenac and naproxen were detected in surface water at different points in the urban stream of Onça, which is located in the city of Três Lagoas, Mato Grosso do Sul, Brazil (Américo et al., 2017). The highest concentration of diclofenac was detected downstream of the discharge of the wastewater treatment plant that serves the city, reaching the highest value of 8.2 μg L^{-1} in late spring. For naproxen, the maximum concentration of 1.8 μg L^{-1} was recorded at the mouth of the stream in early summer.

In the southeast region of the Upper Paraná River Basin, studies were carried out on organic contaminants in the Iguazú River. As mentioned earlier, this river is highly polluted by urban discharges of domestic sewage and industrial wastewater from the city of Curitiba. The concentrations of polynuclear aromatic hydrocarbons (PAHs) in the river are elevated (Table 7.1) and the PAH compounds, chrysene and benzo[a]anthracene were detected at levels higher than the regulatory limits, which are 0.05 μg L^{-1} for PAHs (de Andrade Brito et al., 2018). In addition, Duarte Kramer et al. (2015) reported the occurrence of pharmaceuticals like ibuprofen, diclofenac and paracetamol (or acetaminophen) in the Iguazú River. Ibuprofen was detected only in surface water and at a low frequency of detection, while the other two compounds were present in 100% of the samples. Diclofenac reached values of 285 ng L^{-1} in surface waters while paracetamol was also detected in sediments at 3.74 μg g^{-1}. A very recent study conducted on the bioaccumulation of contaminants by fishes sampled close to the discharge of a municipal wastewater treatment plant into the Paraná River at the city of Posadas in Argentina revealed that there are discharges of microcontaminants of wastewater origin, including caffeine and several pharmaceuticals, such as norfluoxetine, erythromycin, sulfamethoxazole, sulfathiazole and trimethoprim (Ondarza et al., 2019). The last three pharmaceuticals are antibiotics that are used for therapies in human medicine, aquaculture and livestock.

In the Middle and Lower reaches of the Paraná River, several organochlorine pesticides, as well as current use insecticides, cypermethrin and chlorpyrifos and the herbicide, glyphosate have been detected in surface waters (Lenardón et al., 1984; Jergentz et al., 2005; Ronco et al., 2016), as summarized in Table 7.1. In a recent study of the impacts of insecticides in soy production regions of the Pampas in Argentina, Hunt et al. (2017) showed that exposures to mixtures of insecticides can cumulatively reduce the abundance of aquatic benthic invertebrates. Although monitoring studies for pesticides in fishes in the Paraná River are not common, results obtained with migratory species show the potential of using fish for biomonitoring (Ernst et al., 2018; Pérez-Parada et al., 2018). In the Uruguay River, 30 pesticide residues were detected in fish, including migratory and non-migratory species. Pesticides were found in 96% of the 143 samples. A total of 21 pesticides were detected in a single individual *Salminus brasiliensis*, a fish species that includes in its migratory route the Paraná River. Other migratory species that move between the Paraná and Uruguay rivers, such as *Prochilodus lineatus* and *Megaleprinus obtusidens* also incorporated pesticides in their tissues (Ernst et al., 2018). These results indicate the potential contamination of the Paraná River fishes with residues of pesticides. In addition to the lack of monitoring of pesticides in fish, throughout the region, there are few studies that assess the effect of chronic exposure in fish to multi-residues of pesticides. These findings indicate the value of implementing pesticide monitoring programs for fish and assessments of chronic effects in these species in the Paraná River Basin.

Conclusions

The inputs of sediments and nutrients into the Paraná River Basin are highly variable, depending on the area, flood events and climatic cycles. In the Upper Paraná stretch a significant retention of sediments by the network of hydroelectric dams takes place, while nutrients are trapped in the Middle and Lower Paraná stretches by floodplain lakes and the delta of the river, respectively. In addition, the variability in the inputs from anthropogenic sources of pollution and temporal differences between the various studies make it difficult to establish a consistent range of nutrient concentrations throughout the Paraná River.

The basin is affected by anthropogenic sources of pollution and therefore the water quality of the river is poor in many areas. In the Upper Paraná, where hydroelectric dams are present in all major tributaries, the reservoirs contribute to the high nutrient loads that lead to eutrophication and toxigenic cyanobacteria blooms. In the Middle and Lower stretches of the Paraná, discharges of untreated or partially treated domestic wastewater and industrial activities are the main contributors to the deteriorations of water quality.

Metals are primarily sequestered in sediments located in deposition zones in the river, but resuspension of contaminated sediments can lead to downstream transport of these contaminants. Metals can originate from geological sources, but there are significant inputs of metals into the Paraná River Basin from domestic wastewater and industrial effluents.

There are inputs of agrochemicals throughout the basin as a result of intensive crop production (e.g. sugarcane, soybeans). Current use pesticides, including the herbicides glyphosate and metolachlor, as well as organophosphate, pyrethroid and neonicotinoid insecticides and some fungicides have been detected throughout the basin, including waters in places used for human consumption without treatment. Legacy compounds like the persistent organochlorine insecticide endosulfan and its metabolite endosulfan sulfate are still ubiquitous, despite the ban on use in 2013.

Several information gaps emerge from this review of the literature, mainly in terms of scarce integrated monitoring programs at regional levels along the river and of the lack of use of modern techniques for the early detection of pollution (e.g. biomarkers). One reason for that could be that the governance of the Paraná River is highly fragmented and in the hands of multiple bilateral and multilateral legal instruments, a fact that makes it difficult to develop monitoring programs for water quality along the entire river.

Acknowledgements

Many thanks to Dieison André Moi, Florencia Gutierrez, Simoni Loverde, Roger Mormul, Chris Metcalfe and Pablo Collins for helping to locate some of the literature on water quality for the Paraná River Basin. Thanks to Dieison André Moi for providing the lead photo by authorization from Itaipú Binacional. Thank you very much to the reviewers that improved the chapter with their comments.

References

Abarca, F.J. 2007. Tecnícas para evaluación y monitoreo del estado de humedales y otros ecosistemas acuáticos. In: Sanchez, O., Herzig, M., Peters, E., Márquez, R., Zambrano, L. (eds.), *Perspectivas sobre conservación de ecosistemas acuáticos en México. Secretaría de Medio Ambiente y Recursos Naturales*. Instituto Nacional de Ecología, México, pp. 113–144.

Agostinho, A.A., Costa Bonecker, C., Gomes, L.C. 2009. Effects of water quantity on connectivity: The case of the upper Paraná River floodplain. *Ecohydrology and Hydrobiology* 9: 99–113.

Agostinho, A.A., Gomes, C.L., Suzuki, H.I., Ferreira Júlio, H., Jr. 2003. Migratory fishes of the Upper Paraná River basin, Brazil. In: Carosfeld, J., Harvey, B., Ross, C.V., Baer, A. (eds.), *Migratory Fishes of South America: Biology, Fisheries and Conservation Status*, International Development and Research Centre, Ottawa, ON, Canada, pp. 23–98.

Agostinho, A.A., Thomaz, S.M., Minte- Vera, C.V., Winemiller, K.O. 2000. Biodiversity in the high Paraná river floodplain. In: Gopal, B., Junk, W.J., Davis, J.A. (eds.), *Biodiversity in Wetlands: Assessment, Function and Conservation*, Backhuys Publishers, Leiden, The Netherlands, vol. 1, pp. 89–118.

Américo, J.H.P., Isique, W.D., Hortense Torres, N., Machado, A.A., de Carvalho, S.L., Filho, W.V.V., Romanholo Ferreira, L.F. 2017. Ocorrência de diclofenaco e naproxeno em agua superficial no município de Três Lagoas (MS) e a influência da temperatura da água na detecção desses anti-inflamatórios. *Engineiro Sanitacion Ambiental* 22:429–435.

Barros, I.T., Ceccon, J.P., Glinski, A., Liebel, S., Grötzner, S.R., Randi, M.A.F., Benedito, E., Ortolani-Machado, C.F., Neto, F.F., de Oliveira Ribeiro, C.A. 2017. Environmental risk assessment in five rivers of Paraná River basin, Southern Brazil, through biomarkers in *Astyanax spp. Environmental Science and Pollution Research* 24:16228–16240.

Batista de Figueiredo, S. 2012. Avaliacao da qualidade da agua da sub-bacia do Rio Cuiabá-MT aplicando análise multivariada. Doctoral dissertation, Universidade Federal de Mato Grosso, Brasil, p. 140 p.

Bergier, I., Assine, M.L. 2016. Dynamics of the Pantanal wetland in South America. In: Barceló, D., Kostianoy, A.G. (eds.), *The Handbook of Environmental Chemistry 37, Series Editors*, Springer, Geneva, Switzerland, p. 106.

Bertón, A. 2017. *Amostragem instantanea e passive na análise de pesticides em agua de nascente do alto do Rio São Lourenço em Campo Verde-MT*. Doctoral dissertation, Universidade Federal de Mato Grosso- Brasil, p. 89.

Blettler, M.C.M., Oberholster, P.J., Madlala, T., Eberle, E.G., Amsler, M.L., De Klerk, A.R., Truter, J.C., Marchese, M.R., Latosinski, F.G., Szupiany, R. 2018. Habitat characteristics, hydrology and anthropogenic pollution as important factors for distribution of biota in the middle Paraná River, Argentina. *Ecohydrology and Hydrobiology*. DOI:10.1016/j.ecohyd.2018.08.002.

Bonetto, A.A. 1986. The Paraná River system. In: Davies, B.R., Walker, K.F. (eds.), *The Ecology of River Systems*, Springer Science+Business Media Dordrecht, Dordrecht, pp. 541–556.

Bonetto, C., de Cabo, L., Gabellone, N., Vinocur, A., Donadelli, J., Unrein, F. 1994. Nutrient dynamics in the deltaic floodplain of the Lower Paraná River. *Archiv fur Hydrobiologie* 131:277–295.

Brugnoli, E., Muniz, P., Venturini, N., Brena, B., Rodríguez, A., García-Rodríguez, F. 2019. Assessing multimetric trophic state variability during an ENSO event in a large estuary (Río de la Plata, South America). *Regional Studies in Marine Science* 28:565–576.

Burton, G.A., Jr. 2002. Sediment quality criteria in use around the world. *Limnology* 3:65–75.

Campodonico, V.A., García, M.G., Pasquini, A.I. 2015. The dissolved chemical and isotopic signature downflow of the confluence of two large rivers: The case of the Paraná and Paraguay rivers. *Journal of Hydrology* 528:161–176.

Carmichael, W.W., Boyer, G.L. 2016. Health impacts of cyanobacteria harmful algal blooms: Implications for the North American Great Lakes. *Harmful Algae* 54:194–212.

Cataldo, D., Colombo, J.C., Boltovskoy, D., Bilos, C., Landoni, P. 2001. Environmental toxicity assessment in the Paraná river delta (Argentina): Simultaneous evaluation of selected pollutants and mortality rates of *Corbicula fluminea* (Bivalvia) early juveniles. *Environmental Pollution* 112:379–389.

Clemente, Z., Busato, R.H., Oliveiro Ribeiro, C.A., Cestari, M.M., Ramsdorf, W.A., Magalhães, V.F., Wosiack, A.C., Dilva de Assis, H.C. 2010. Analyses of paralytic shellfish toxins and biomarkers in a southern Brazilian reservoir. *Toxicon* 55:396–406.

de Andrade Brito, I., Esquivel Garcia, J.R., Barbosa Salaroli, A., Lopes Figueira, R.C., de Castro Martins, C., Cordeiro Neto, A., Kachel Gusso-Choueri, P., Brasil Choueri, R., Borges Lino Araujo, S., de Oliveira Ribeiro, C.A. 2018. Embryotoxicity assay in the fish species *Rhamdia quelen* (Teleostei, Heptaridae) to assess water quality in the Upper Iguaçu basin (Parana, Brazil). *Chemosphere* 208:207–218.

Depetris, P.J., Kempe, S. 1993. Carbon dynamics and sources in the Paraná River. *Limnology and Oceanography* 387:382–395.

Depetris, P.J., Pasquini, A.I. 2007. The geochemistry of the Paraná River: An overview. In: Iriondo, M.H., Paggi, J.C., Parma, M.J. (eds.), *The Middle Paraná River. Limnology of a Subtropical Wetland*, Springer, Berlin, pp. 143–174.

Desarrollos de Niveles Guía Nacionales. 2003. Desarrollos de niveles guía nacionales de calidad de agua ambiente correspondientes a Cromo. Subsecretaría de Recursos Hídricos de la Nación, p. 42.

Dos Santos Rosa, R., Fornero Aguiar, A.C., Gonçalves Boëchat, I., Gücker, B. 2013. Impacts of fish farm pollution on ecosystem structure and function of tropical headwater streams. *Environmental Pollution* 174:204–213.

Drago, E.C., Amsler, M.L. 1998. Bed sediment characteristics in the Paraná and Paraguay rivers. *Water International* 23:174–183.

Duarte Kramer, R., Mizukawa, A., Honjo Ide, A., Oliveira Marcante, L., Marques Dos Santos, M., Rodrigues de Azevedo, J.C. 2015. Determinação de anti-inflamatórios na água e sedimento e suas relações com a qualidade da água na bacia do Alto Iguaçu, Curitiba-PR. *Revistas Brasilieros de Recursos Hídricos* 20:657–667.

Ernst, F., Alonso, B., Colazzo, M., Pareja, L., Cesio, V., Pereira, A., Márquez, A., Errico, E., Marley, A., Heinzen, H., Pérez-Parada, A. 2018. Occurrence of pesticide residues in fish from south American rainfed agroecosystems. *Science of the Total Environment* 631:169–179.

Etchegoyen, M.A., Ronco, A.E., Almada, P., Abelando, M., Marino, D.J. 2017. Occurrence and fate of pesticides in the Argentine stretch of the Paraguay-Paraná basin. *Environmental Monitoring and Assessment* 189:63.

Facetti-Masulli, J.F., Klump, P. 2010. Selected minor and trace elements from water bodies of western Paraguay. *Journal of Radioanalytical Nuclear Chemistry* 286:441–448.

Frata, A.M., Batolomeu de Faria, A. 2010. A expansao da cana-de-acúcar e o recorte em bacias hidrográficas: O etanol na sub-bacia do Rio Ivinhema. *48 Congreso SOBER. Campo Grande*, July 25–28.

Furtado Gonçalves, M., Tiemy Kishi, R., Scapulatempo Fernandes, C.V. 2016. Poluição por metais pesados na bacia do rio Barigui, estado do Paraná. *REGA* 13:175–185.

Gallo, M., Trento, A., Alvarez, A., Beldoménico, H., Campagnoli, D. 2006. Dissolved and particulate heavy metals in the Salado River (Santa Fe, Argentina). *Water, Air, and Soil Pollution* 174:367–384.

Gonino, G.M., Figueiredo, B.R., Manetta, G.I., Alves, G.H.Z., Benedito, E. 2019. Fire increases the productivity of sugarcane, but it also generates ashes that negatively affect native fish species in aquatic systems. *Science of the Total Environment* 664:215–221.

Hunt, L., Bonetto, C., Marrochi, N., Scalise, A., Fanelli, S., Liess, M., Lydy, M.J., Chiu, M.-C., Resh, V.H. 2017. Species at Risk (SPEAR) index indicates effects of insecticides on stream invertebrate communities in soy production regions of the Argentine Pampas. *Science of the Total Environment* 580:699–709.

Hunt, L., Bonetto, C., Resh, V.H., Forsin Buss, D., Fanelli, S., Marrochi, N., Lydy, M.J. 2016. Insecticide concentrations in stream sediments of soy production regions of South America. *Science of the Total Environment* 547:114–124.

Iriondo, M. 2004. The littoral complex at the Paraná mouth. *Quaternary Internat.* 114:143–154.

Isupova, M.V., Mikhailov, V.N. 2018. Long-term variations of water runoff and suspended sediment yield in the Paraná and Uruguay rivers. *Water Resources* 45:846–860.

Jergentz, S., Mugni, H., Bonetto, C., Schulz, R. 2005. Assessment of insecticide contamination in runoff and stream water of small agricultural streams in the main soybean area of Argentina. *Chemosphere* 61:817–826.

Laabs, V., Amelung, W., Pinto, A.A., Wantzen, M., da Silva, C.J., Zech, W. 2002. Pesticides in surface water, sediment, and rainfall of the northeastern Pantanal basin, Brazil. *Journal of Environmental Quality* 31:1636–1648.

Larson, C.A., Merritt, G., Janisch, J., Lemmon, J., Rosewood-Thurman, M., Engenesse, B., Polkowske, S., Onwumere, G. 2019. The first statewide stream macroinvertebrate bioassessment in Washington State with a relative risk and attributable risk analysis for multiple stressors. *Ecological Indicators* 102:175–185.

Latrubesse, E.M., Abad, J.D. 2018. Preface to the volume, Large Rivers. *Geomorphology* 302:1–2.

Le Moal, M., Gascuel-Odoux, C., Ménesguen, A., Souchon, Y., Etrillard, C., Levain, A., Moatar, F., Pannard, A., Souchu, P., Lefebvre, A., Pinay, G. 2019. Eutrophication: A new wine in an old bottle? *Science of the Total Environment* 651:1–11.

Lenardón, A.M., de Hevia, M.I.M., Fuse, J.A., de Nochetto, C.B., Depetris, P.J. 1984. Organochlorine and organophosphorous pesticides in the Paraná River (Argentina). *Science of the Total Environment* 34:289–297.

Maine, M.A., Suñe, N.L., Bonetto, C. 2004. Nutrient concentrations in the Middle Paraná River: Effect of the floodplain lakes. *Archives of Hydrobiology* 160:85–103.

Mesa, L., Mayora, G., Saigo, M., Giri, F. 2015. Nutrient dynamics in wetlands of the Middle Paraná River subjected to rotational cattle management. *Wetlands* 35:1117–1125.

Milliman, J.D., Farnsworth, K.L., 2011. *River Discharge to the Coastal Ocean. A Global Synthesis*. Cambridge University Press, Cambridge, UK.

Natale, O.E. 2005. *Water Quality Indicators for the La Plata River Basin*. IWG-Env, International Work Session on Water Statistics, Vienna, June 20–22.

Neto, R.M., Nocko, H.R., Ostrensky, A. 2017. Carrying capacity and potential environmental impact of fish farming in the cascade reservoirs of the Paranapanema River, Brazil. *Aquaculture Research* 48:3433–3449.

Ondarza, P.M., Haddad, S.P., Avigliano, E., Miglioranza, K.S.B., Brooks, B.W. 2019. Pharmaceuticals, illicit drugs and their metabolites in fish from Argentina: Implications for protected areas influenced by urbanization. *Science of the Total Environment* 649:1029–1037.

Orfeo, O., Iriondo, M.H., 2012. Variación quincenal de la concentración y tamaño de sólidos suspendidos del río Paraná (Corrientes, Argentina). *Proceedings of the Meeting on Scientific and Technological Communication*, Universidad Nacional del Nordeste, Corrientes, Argentina, 118 pp.

Pasquini, A.I., Depetris, P.J. 2012. Hydrochemical considerations and heavy metal variability in the middle Paraná River. *Environmental Earth Science* 65:525–534.

Peluso, L., Abelando, M., Apartín, C.D., Almada, P., Ronco, A.E. 2013. Integrated ecotoxicological assessment of bottom sediments from the Paraná basin, Argentina. *Ecotoxicology and Environmental Safety* 98:179–186.

Peraza-Castro, M., Sauvage, S., Sánchez-Pérez, J.M., Ruiz-Romera, E. 2016. Effect of flood events on transport of suspended sediments, organic matter and particulate metals in a forest watershed in the Basque country (Northern Spain). *Science of the Total Environment* 569–570:784–797.

Pérez-Parada, A., Goyenola, G., Teixeira de Mello, F., Heizen, H. 2018. Recent advances and open questions around pesticide dynamics and effects on freshwater fishes. *Current Opinions in Environmental Science and Health* 4:38–44.

Pinto Martins, D., Bravard, J.-P., Stevaux, J.C. 2009. Dynamics of water flow and sediments in the Upper Paraná River between Porto Primavera and Itaipu dams. *Brazilian Latin American Journal of Sedimentology and Basin Analysis* 16:111–118.

Possavatz, J. 2013. Determinação de resíduos de pesticidas, em amostras de agua e sedimento, na bacia hidrográfica do rio Cuiabá, Mato Grosso. *Masters thesis from the Post-graduate program in aquatic resources*, Universidade Federal Mato Grosso, Cuiabá, Brazil, p. 92.

Ronco, A.E., Marino, D.J.G., Abelando, M., Almada, P., Apartin, C.D. 2016. Water quality of the main tributaries of the Paraná Basin: Glyphosate and AMPA in surface water and bottom sediments. *Environmental Monitoring and Assessment* 188:458.

Sica, Y.V., Quintana, R.D., Radeloff, V.C., Gavier-Pizarro, G.I. 2016. Wetland loss due to land use change in the Lower Paraná River delta, Argentina. *Science of the Total Environment* 568:967–978.

Suárez, P. 2012. *Dinámica de elementos solubles del río Negro (Provincia del Chaco, Argentina) y su relación con los sedimentos fluviales*. PhD Thesis. Repositorio University of Valladolid. Retrieved May 27, 2015. http://uvadoc.uva.es/handle/10324/4881.

UNSD/UNEP. 2004. *Questionnaire on Environment Statistics. Water Section*. Retrieved from: https://unstats.un.org/unsd/envstats/questionnaire.

UN-Water. 2017. Step-by-step monitoring methodology for indicator 6.3.2. *Integrated Monitoring Guide for SDG* 6: 27. Retrieved from: www.unwater.org/publications/step-step-methodology-monitoring-water-quality-6-3-2/. Accessed May 20, 2019.

Veiga, B.V., Dziedzic, M. 2010. Estimating nutrient loads in the Passaúna reservoir with FLUX. *Water International* 35:210–222.

Viana, L.F., Suárez, Y.R., Cardoso, C.A.L., Do Amaral Crispim, B., Grisolia, A.B., Lima-Junior, S.E. 2017. Mutagenic and genotoxic effects and metal contaminations in fish of the Amambai River, Upper Paraná River, Brazil. *Environmental Science and Pollution Research* 24:27104–27112.

Viana, L.F., Suárez, Y.R., Cardoso, C.A.L., Solórzano, J.C.J., Do Amaral Crispim, B., Grisolia, A.B., Lima-Junior, S.E. 2018. Erythrocyte nuclear abnormalities in *Astyanax lacustris* in response to landscape characteristics in two neotropical streams. *Archives of Environmental Contamination and Toxicology* 75:327–334.

Villar, C.A., de Cabo, L., Vaithiyanathan, P., Bonetto, C. 1998. River-floodplain interactions: Nutrient concentrations in the Lower Paraná River. *Archives of Hydrobiology* 142:433–450.

Weneck Lima, J.E., Araújo Lopez, W.T., de Olivera Carvalho, N., Vieira, M.R., Medrado da Silva, E. 2005. Suspended sediment fluxes in the large river basins of Brazil. Sediment Budgets 1, Proceedings of symposium S1 held during the Seventh IAHS Scientific Assembly at Foz do Iguaçu, IAHS Publication 291, Brazil, April 2005.

8

Transboundary cooperation

Alejandro Pastori and Francesco Sindico

Key messages

- The basin of the Paraná River is governed by many different bilateral and multilateral legal instruments.
- A complex web of Commissions linked to the basin of the Paraná River provides the institutional backbone for the governance system for the basin.

- The Rio de la Plata Basin Treaty which was signed in 1969 by Argentina, Bolivia, Brazil, Paraguay and Uruguay, provided the framework for a long list of bilateral and multilateral agreements that have since been adopted to regulate the Paraná River and the entire basin.
- Despite the number of treaties signed to protect surface water resources within the basin, only recently has there been an interest in developing treaties to protect transboundary aquifers. The Guarani Aquifer Treaty was initially ratified by Argentina, Brazil and Uruguay, and in April 2018, Paraguay finally ratified the treaty, opening the door to its entry into force.
- Although the governance of the Paraná River is highly fragmented, there have not been any major instances in the past of conflicts.
- Where tensions exist, these are related to the implementation of energy-related treaties that regulate the sharing of revenues from hydropower in the basin.

Introduction

The Paraná River begins in Brazilian territory, crosses the boundary between Brazil and Paraguay, then between Paraguay and Argentina, and concludes its course in Argentina. The discharge from the Paraná River and the discharge from the Uruguay River both flow into the Rio de la Plata estuary, which then flows into the Atlantic Ocean. Although the Uruguay River is considered a separate hydrological system from the Paraná River, some of the international treaties also include Uruguay, through which the Uruguay River flows.

As discussed in Chapter 1 in this book, the basin of the Paraná River covers an area of approximately 2.6 million km^2. The tributaries along its eastern margin include the headwaters of the Paranaiba and Rio Grande, the Tietê, the Paranapanema, the Verde, the Pardo, the Mondai and the Iguazú. The Paraguay River and its tributaries form the western margin of the basin, which includes the eastern part of Bolivia. Of particular importance to this chapter is the Paraguay River, as the Paraná-Paraguay Waterway is probably the most prominent example of cooperation in the region. The Paraná Basin partly overlies the Guarani Aquifer System, the second largest source of groundwater in the world (Amore, 2011).

The Upper stretch of the Paraná River is characterized by a large and highly permanent flow, large altitude gradient along its course, canyon formations in certain sections and basaltic structures, all of which combined constitute exceptionally favorable conditions for the construction of hydroelectric power plants. It is not surprising then that approximately 140 hydroelectric plants have been built and are currently in operation. Several more dams have been projected for construction in the area. The Yaciretá hydroelectric dam on the Upper Basin of the Paraná River was constructed through a cooperative agreement between Paraguay and Argentina (Figure 8.1).

Because the Paraná River and the rest of the basin run through the four countries of Brazil, Bolivia, Paraguay and Argentina, much of their administration operates bi-nationally or multilaterally and this has led to the adoption of several treaties, the concurrent application of which may seem quite confusing. However, somehow, the Paraná governance structure works through a web of formal and informal cooperative agreements among State governments and technical agencies. This chapter first provides an analysis of the bilateral and multilateral legal instruments that provide the backbone of the governance structure for the Paraná River and the basin. It then explores the institutions that have been developed by such legal instruments and that deliver the management of the river and the basin. The chapter will then move on to discuss specific examples of cooperation between some of these institutions before exploring the potential for conflict in what is indeed a very fragmented landscape. The chapter concludes by providing

Figure 8.1 The Yaciretá dam shared by Argentina and Paraguay on the Upper Paraná River.
The photograph was provided by agreement with commons.wikimedia.org.

arguments that the Paraná is managed effectively, and conflict has been avoided between riparian countries, despite the existing fragmentation and some imbalance in hydropower usage.

The legal framework for cooperation

A plethora of international agreements signed over the 20th century govern the Paraná basin, including the Paraná River, its tributaries and the activities carried out along its course, as well as the aquifer that lies underneath. In this first section, we analyze the legal framework that regulates the use of water in the Paraná basin. These legal agreements and the institutions that have been created to administer the agreements are summarized in Table 8.1. The state of cooperation within the basin has been reviewed previously (Gilman et al., 2008; Laborde, 2008; Villar et al., 2018). As we will see below, this legal framework is composed of several treaties, which tend to give the impression of a decentralized system. However, the four countries involved (i.e. Brazil, Bolivia, Paraguay, Argentina) plus Uruguay have created throughout the years a series of rules and criteria drafted in meetings of experts on the use of water, which were later incorporated into bilateral treaties signed by the Cooperating States in order to manage together the use of the water resources of the Paraná Basin. The primary treaties are reviewed over the remainder of this chapter.

The framework treaty: Rio de la Plata basin treaty (Brasilia, April 23, 1969)

This treaty, which was signed by Argentina, Bolivia, Brazil, Paraguay and Uruguay, sought to ensure '*joint action, allowing for the harmonious and balanced development of the region, as well as the optimal use of its vast natural resources, ensuring their preservation for future generations through the rational use of such resources*'.[1] In practice, the purpose of the treaty was to establish a framework that would allow the States to reach operational agreements and sign other, more specific, legal instruments for the development of activities in the basin.[2]

Although the Rio de la Plata Basin Treaty itself does not include substantive rules about the use of watercourses, the First Meeting of the Ministers of Foreign Affairs of the States of the Rio de la Plata Basin, held on June 3, 1971 passed Resolution 25, which adopted the Declaration of Asunción on the Use of International Rivers. This Declaration states the basic criteria for the development and use of water resources in the area, such as the need for bilateral agreements to regulate the use of international rivers; the principle of cooperation; the principle of reasonable and equitable use; the obligation to maintain the navigability of watercourses; and the obligation to protect and preserve living resources.[3]

In accordance with the mandate of the Rio de la Plata Basin Treaty and in keeping with the spirit of Resolution 25, a long list of bilateral and multilateral agreements has since been adopted. These include the treaties that regulate the Paraná River and the entire basin as these contribute the majority of the flows within the Rio de la Plata Basin.

The navigation and trade treaty: the Paraná-Paraguay Waterway treaty (June 26, 1992)

The Paraná-Paraguay Waterway project was established in the 1969 Rio de la Plata Basin Treaty, but it was not until 1987 in a Meeting of the Ministers of Foreign Affairs held within the framework of that Treaty in Santa Cruz de la Sierra that a Resolution was signed collectively that defined navigation and port development as a top priority.[4]

A year later in 1988, the Ministers of Public Works and Transportation met in Campo Grande, Brazil, to hold the First International Meeting for the Development of the Parana-Paraguay Waterway, in order to determine the best options for the development of the waterway, as a means of regional transportation and as the backbone for future integration.

In 1989, a milestone was reached when the presidents of the five countries signed a more specific declaration. The 19th Meeting of the Ministers Foreign Affairs of States Parties to the River Plate Basin Treaty adopted Resolution No. 239, seeking the improvement of the waterway (making it navigable 24 hours per day), the development of ports, the enhancement of the fleet and the harmonization of national navigation regulations.[5] Following this declaration, the Ministers of Transport and Public Works created the Intergovernmental Waterway Committee (i.e. Comité Intergubernamental de la Hidrovía; CIH), which was aimed at identifying and prioritizing specific projects and developing a unified body of navigation regulations.

A singularity revolves around this Waterway agreement. Although it was originally a vision that stemmed from the Rio de la Plata Basin Treaty, the parallel efforts to encourage the economic integration of the five State Parties led to the celebration of the Waterway Treaty under a different framework. In fact, the Waterway Treaty was celebrated under the Montevideo Treaty of 1980 that created the Latin American Integration Association (ALADI), and is, to this date, a Partial Scope Agreement registered before this organization, based on the principles, object, mechanisms and resolutions of ALADI.[6] Consequently, as a Partial Scope Treaty of ALADI, the Waterway Agreement is an autonomous treaty, independent in its activities and decisions from both ALADI and the Intergovernmental Committee of the Rio de la Plata Basin, despite its evident connection to both organizations.

Article 1 of the Waterway Agreement provides that its object is to '*facilitate navigation and commercial transportation through the length of the river (...) establishing a legal framework that will favor development, modernization, and efficiency in operations, as well as allow all Parties access in competitive circumstances to markets overseas*'. In order to achieve these goals, the Agreement covers navigation, trade and transportation of goods and persons, with the exception of war ships and other not-for-profit vessels. Navigation across the riverbanks is also excluded from the regulations of the Waterway Agreement.

In general, the Waterway Agreement reiterates that ships bearing the flag of any of the State Parties shall have the right of free passage through the length of the Waterway and shall grant the same right to ships of third parties. Persons and goods shall transit freely through the Waterway, with the sole obligation to pay such fees as may be charged for services rendered, including loading and unloading, cargo transfer and deposit of merchandise, among others. Although significant progress has been made towards adopting a common regulatory framework for the Waterway, and generating unified technical specifications, safety and economic rules, not all of the State Parties have internalized these regulations. Another international agency was created by the Waterway Agreement to regulate the activities and relations of the Parties; the above mentioned CIH.

Starting in 1992, the CIH has generated several sets of specific rules that govern issues such as customs, safety, insurance, equalization of taxes, technical matters, pilotage, ship security and staff training, among others. While all of these rules have been adopted by Bolivia, Brazil and Uruguay, the other two State Parties, Argentina and Paraguay, have only adopted some, and have been reluctant to agree to others.

The Waterway Agreement has been in force since 1995.[7] Although its original text established a duration of ten years, the State Parties signed its Seventh Additional Protocol in 2004, extending the duration of the Agreement for another 15 years, as of February 13, 2005. The current term of the Agreement will expire in February of 2020.

The transboundary treaty: agreement between Argentina and Paraguay to study the use of the resources of the Paraná River

The geographical area covered by this treaty, hereinafter referred to as the Paraná River Treaty, is limited to the stretch that forms the boundary between Paraguay and Argentina. In order to govern this area, Paraguay and Argentina signed a Treaty in 1971, creating the Argentinian-Paraguayan Mixed Commission of the Paraná River (Comisión Mixta Argentino-Paraguaya del Río Paraná; COMIP).[8] The object and purpose of the Commission was to conduct '*studies and assessments regarding the technical and economic feasibility of using the resources of the Parana River*'.[9] However, there is one feature that the treaty expressly excludes, which is the regulation and operation of the Yacyretá-Apipe Dam and Hydroelectric Plant, a pre-existing project that is governed by a different agreement between the State Parties.

The power generation agreements: Itaipu and Yacyretá treaties

As stated in the introduction to this chapter, one of the Paraná Basin's richest resources is water flow, and the incredible amount of power generation that it brings. Certainly, States have built and operated their own hydropower plants, but some of these plants are operated by two State Parties to the River Plate Basin Treaty, which has created the need to sign bilateral and multilateral agreements. The two largest hydropower plants in the Paraná Basin area are the Itaipu and Yaciretá. Their history is far from seamless, since several years and rounds of negotiations were needed for all parties affected to sign off on their construction, and even today there are serious claims as to who benefits from their operation.

Background

The Itaipu dam was constructed in a part of the river where only Paraguay and Brazil have jurisdiction. Argentina made an early claim that the change in the flow of the Paraná River would negatively affect the generation of power at its Corpus Christi plant, which was a project downstream, between Argentina and Paraguay. Argentina reserved the right to request the Brazilian and Paraguayan governments to revise and make modifications to any project that might alter the possibilities for exploitation of hydropower and, in general, any prejudicial change in the natural course of the Paraná River (Querol, 2002), in accordance with the doctrine of absolute territorial integrity (Aguilar Rojas and Iza, 2009).

The concern raised by Argentina was a valid one, since States are obliged to respect the correlative rights of other States over shared water resources (McCaffrey, 2007). The conflict was resolved by the signing of yet another multilateral agreement between Argentina, Brazil and Paraguay on October 19, 1979; the Tripartite Technical and Operational Cooperation Agreement between Itaipu and Corpus Christi.[10] In this new treaty, the State Parties agreed to adopt all measures necessary to maintain the best possible navigability conditions in the rivers subject to their jurisdiction, taking into account the interests of the riparian States downstream and upstream. The three States also agreed to make all efforts to preserve the environment, the fauna, the flora and the quality of the river, avoiding its pollution and ensuring, at least the preservation of the river's conditions in the areas influenced by the parties as of the date of signing of the treaty (Querol, 2002). Significantly, the Corpus Christi dam is still just a planned project, while the Itaipu and Yaciretá are today among the largest hydropower generating complexes in the world.

Treaty of Itaipu (signed August 13, 1973)

The Itaipu hydropower plant is located on the border between Brazil and Paraguay (Figure 8.2). It was built upon the celebration of the Treaty between the Federative Republic of Brazil and the Republic of Paraguay for the Hydropower Exploitation of Water Resources of the Paraná River, which is a Condominium of both States (Treaty of Itaipu). With 20 generating units, and 14,000 MW of installed power, Itaipu provides about 16% of the energy for Brazil, and 86% of the energy for Paraguay.[11] The Treaty of Itaipu is a hydropower generation agreement developed for the sole purpose of regulating the use of the Paraná River for power generation.[12] As established in its Article 1, integral to this agreement are the three Annexes, which regulate the creation and Statute of the ITAIPU Binational Entity (Annex A).[13]

A general description of the main and auxiliary facilities that will be used for the production of electric power (Annex B) and the Financial and Power Services bases of the Itaipu plant (Annex C). The Treaty provides that the power generated shall be divided in equal shares between both State Parties, and each shall be recognized as having a right to purchase any portion of power that is not utilized by the other party (Sorgine, 2012). However, the entire amount of power generated by the plant must be purchased by the Parties of the Itaipu Treaty. In practice, this means that the ITAIPU Binational Entity pays royalties to both Brazil and Paraguay in equal shares for the use of the power generated by the plant. However, it also

Figure 8.2 The Itaipu dam and reservoir located on the Upper Paraná River.
The photograph was provided by Dieison André Moi by authorization from Itaipú Binacional.

means that there is a prohibition to sell energy to any third State, even if one of the parties does not use its share entirely. Paraguay covers nearly 86% of its national power needs with a small fraction of the energy produced by Itaipu (28%).[14] Therefore, every year it is left with a large unused portion of electrical power that it is obligated to sell to Brazil, at prices well below market value, which has led Paraguay to claim that it should have the right for fair payment from Brazil for the energy produced by Itaipu, or the chance to sell to third parties.[15]

Treaty of Yaciretá (signed December 3, 1973)

Another major power generating complex in the Paraná River is the Hydropower Plant of Yaciretá-Apipe. With an installed generating power of 3,200 MW h^{-1}, Yaciretá currently supplies 22% of the power demands of Argentina, representing 45% of the total hydroelectric power generating capacity of the country. This dam was built upon the celebration of the Treaty of Yaciretá between Argentina and Paraguay.[16]

The project to build Yaciretá was agreed upon in 1958, with the creation of an Argentine-Paraguayan Mixed Commission that was charged with carrying out all the technical studies necessary for the exploitation of electrical power and improvements to navigation conditions in the Paraná River in and around the area of the Yaciretá and Apipe Islands.[17] The original Bilateral Technical Commission that had been created in 1958 was included in the final Treaty of Yaciretá,[18] and currently oversees the operation of the hydropower plant. The structure of the Treaty of Yaciretá is identical to that of the Treaty of Itaipu, and it also includes the same Annexes described above.

As well as regulating the operations of the hydropower plant, the Treaty of Yaciretá also governs the efforts of the two states to minimize the damaging effects of floods, by creating a flood alert system and facilitating the implementation of irrigation systems. Another important difference is that the dam in Yaciretá includes a sluicegate system that guarantees free navigation. This is a concrete structure that forms a lock 27 m wide and 236 m tall, with a sluicegate at the top and another at the bottom.

As established in the Treaty of Yaciretá, the terms for the distribution of power generated by the plant between Argentina and Paraguay are quite similar to those of the Itaipu Treaty (Newman, 2010). Each State Party has a right to use 50% of the generated power, and all the remaining energy generated by the plant must be purchased by one of the Parties to the treaty. Consequently, the effects of these terms are also similar to the Itaipu case, as there is a *de facto* prohibition to sell energy to any third party, and therefore the State with the surplus must sell to the other treaty party at prices well below market value. In fact, the Treaty of Yaciretá contains wording that is even more severe than that of Itaipu, as it does not refer to a 'sale' of the surplus energy, but rather uses the term 'cession' (Article XV, paragraph 3), and instead of 'payment', there will be 'compensation'. The practical effect is that there is really no equal sharing of the benefits derived from exploiting the Paraná River:

> *Our partners in this equal-rights condominium of the Yacireta project have taken 94.3% of the energy production in 20 years. Additionally, they have used 88.59% of the Paraguayan share – that is 115,261,972 MWh. This is an extremely significant number when compared to the national* [Paraguayan] *use of the energy, which is around 12,000 MWh per year. The worst part is that this electrical power, which had a regional market value of up to US $500/MWh, was paid at an average of US $9/MWh over cost.*[19]

Suffice to say that Paraguay continues to request a revision of both treaties in order to achieve a more equitable gain from the exploitation of its shared resources. This revision has been in the making for over five years, in the form of an amendment to Annex C of the Treaty of Yaciretá. However, no agreement has yet been achieved.[20]

The groundwater agreement: the Guarani aquifer treaty

A peculiarity of the vast array of treaties that govern the watercourses of the Paraná Basin is that they contain no rules whatsoever regarding the use and protection of groundwater. This is striking when one considers that the Guarani Aquifer is the second largest source of groundwater in the world, with a volume of stored groundwater of between 40 and 50,000 km^3, and an exploitable reserve of 2,000 km^3 per year. The aquifer recharges at a rate of 6 km^3 year^{-1}. These numbers mean that the Guarani Aquifer could provide sustainable fresh water to the entire world population for 200 years, at a usage rate of 100 liters per inhabitant per day. Four States share sovereignty over the aquifer's area of 1.2 million km^2. Approximately, 840,246 km^2 are located in Brazil, 225,424 km^2 in Argentina, 72,540 km^2 in Paraguay and 58,545 km^2 in Uruguay.

By 2010, the country that was using the largest portion of water reserves from the Guarani Aquifer was Brazil, providing water to roughly 400 cities. In the same year, Uruguay had 135 public wells, some of which were used for thermal waters, Paraguay had 200 wells for human use and Argentina had five freshwater wells, and only one of brackish water.[21] Significantly, the negotiations that led to the signing of the Guarani Aquifer Agreement (GAA)[22] were conducted practically at the same time as the 2008 negotiations for the Draft Articles on the Law of Transboundary Aquifers (hereinafter Draft Articles), adopted by Resolution A/RES/63/124 of the UN General Assembly. It follows that the GAA was the first treaty regulating a transboundary aquifer signed after the adoption of the Draft Articles. The GAA is a framework agreement that has established the basic principles for the future administration of the aquifer, based on cooperation, setting the pillars of a future institutional structure (Sindico, 2011; Sindico et al., 2018).

The GAA refers to the countries' national sovereignty, but such reference is balanced with the need to fully conform to international law. States Parties to the GAA are, hence, bound to customary international law and to the Draft Articles when its provisions can be said to reflect

Table 8.1 Legal instruments and institutions created for transboundary cooperation in the management of surface water and groundwater resources in the basin of the Paraná River, listed in chronological order.

Name of the Legal Instrument	*State Parties*	*Institutions Created*	*Year of Adoption*	*Relationship with the Paraná River and Basin*
River Plate Basin Treaty	Argentina, Bolivia, Brazil, Paraguay and Uruguay	CIC	1969	It is the framework agreement under which all subsequent treaties that regulate more specific territories or activities were created.
Paraná River Treaty	Argentina and Paraguay	COMIP	1971	Technical and economic benefits of hydroelectric power generation.
Itaipu Treaty	Paraguay and Brazil	ITAIPU	1973	Hydroelectric power generation.
Yaciretá Treaty	Paraguay and Argentina	EBY	1973	Hydroelectric power generation.
Tripartite Agreement between Itaipu and Corpus Christi	Paraguay, Argentina and Brazil	None	1979	Ensuring the flow of the Paraná River in order not to affect the planned construction of a hydroelectric power generating plant downstream, at Corpus Christi.
Waterway Agreement	Argentina, Bolivia, Brazil, Paraguay and Uruguay	CIH	1995	It regulates all activities and obligations of the States involved in the Waterway, which covers the southern half of the Paraná River.
Guarani Aquifer Agreement	Brazil, Argentina, Paraguay and Uruguay	Aquifer Commission	2010	The Paraná Basin partly overlies the Guarani Aquifer System, the second largest source of groundwater in the world.

customary international law. Amongst the provisions that apply to the GAA Parties, we can mention the provisions that apply to equitable and reasonable use, the no significant harm, the obligation to notify neighbors of any activities that may cause effects on the aquifer beyond their own borders, and the obligation to exchange information about the state of its resources.[23]

Despite the importance of this transboundary aquifer, until 2018 only Argentina, Brazil and Uruguay had ratified the GAA. Paraguay finally ratified the treaty on 19 April 2018, opening the door to its entry into force.

The surrounding legal framework: other international agreements associated with the Paraná basin

Once again, it should be noted that the Paraná basin is part of a larger system of watercourses. The second major waterway in the basin is the Paraguay River, which originates in the Pantanal region of Brazil in the north, is fed by tributaries originating to the west in the Andean region of Bolivia, and flows south into the Paraná River. Therefore, Bolivia is included in the mix of states that share sovereignty over the Paraná River. All five States are parties together to the Waterway Agreement and the Rio de la Plata Basin Treaty.

However, there are also other bilateral and multilateral agreements that govern the Paraguay River and its main tributaries, the Bermejo River (Treaty between Argentina and Bolivia), the Apa River (Treaty between Brazil and Paraguay) and the Pilcomayo River (Treaty between Argentina, Bolivia and Paraguay).[24] Because the Uruguay River also flows into the Rio de la Plata, Uruguay is also a party to two bilateral treaties with Argentina, which separately govern the use of the Uruguay River and the Rio de la Plata. While these international instruments are not strictly part of the legal framework of the Paraná River Basin, as explained in the next section, the institutions created by all of these treaties interact in one way or another, creating a web of cooperation among the five States involved.

The institutions involved in cooperation

As mentioned before, there is a framework agreement that predates most of the bilateral and multilateral treaties that govern the different watercourses of the Rio de la Plata Basin. One may be tempted to think that the regulation of the entire system would be tied to the Rio de la Plata Basin Treaty. Especially when its object was to '*strengthen the institutionalization of the Basin's system, and coordinate efforts to promote economic development and physical integration* …'[25] It would appear to be both administratively and procedurally more efficient to simply have the Rio de la Plata Basin Treaty's existing institutions cover the oversight of the specific activities and areas covered by any subsequent agreement.

However, the treaties that were celebrated expressly or implicitly within the legal framework created by the Rio de la Plata Basin Treaty went on to create their own institutions and organs. It might not be out of line to think that the creation of 14 different commissions to regulate uses and territory that were already covered by the framework treaty is a huge redundancy. However, a more detailed examination of how this set of treaties and institutions work in practice could explain why there has been such a proliferation of commissions and agencies. This section describes each of the institutions or organs created by the treaties that regulate the use of water resources in the region, starting with the already mentioned framework treaty.

The Rio de la Plata Basin Treaty created an institutional structure to allow its State Parties to do just that; the International Coordinating Committee (CIC, by its Spanish acronym). The legal nature of the CIC as an independent international organization was subject to discussion, as evidenced in the preparatory works of the Rio de la Plata Basin Treaty.[26] However, the legal position of the committee is aptly captured by the statements of Rama Montaldo (1972), who submitted that a detailed examination of the institutional structure created by the treaty reveals that it is '*an international organization created by a permanent association of States, which has organs empowered to express its independent will, and treaty-defined and goals to achieve in exercise of its functions or powers*'.

It should then follow that all activities that may influence the area regulated by this framework agreement should be coordinated by the CIC. However, this has not exactly been the case. The CIC was not the only organ created by the Rio de la Plata Basin Treaty. The actual decision making structure is the Meeting of Foreign Ministers, which should convene once a year on a date suggested by the CIC to determine basic directives of common policy for the achievement of the purpose of the treaty.[27] Hence, the CIC is merely an administrative and technical organ, entrusted to carry out the directives of the State Parties to the treaty, but without any true autonomy. The CIC has an ample array of technical and administrative functions, but it is not the only institution that works in the basin area.

When the Waterway Agreement was drafted, the same member States that were already parties to the River Plate Basin Treaty created two additional institutional structures; the Intergovernmental Waterway Committee (Comité Intergubernamental de la Hidrovía, CIH), with the power to act as the political organ, and the Agreement's Commission, which acts as its technical organ. As a reminder, the Waterway Agreement was not created within the framework of the River Plate Basin Treaty, but under the ALADI framework, as a Partial Scope Agreement.

However, the agreement itself states that the CIH would be an organ of the Rio de la Plata Basin Treaty. In fact, the CIH had existed even before the Waterway Agreement came into force, having been created in 1989.[28] It was integrated into the Rio de la Plata Basin Treaty system at the 20th Meeting of the Foreign Ministers of the Rio de la Plata Basin States.[29] This arrangement leaves the basin with two administrative agencies: the CIC, which oversees all activities in the basin, and the CIH, which was tasked with managing the Waterway and all activities derived from that particular use of the Paraná and Paraguay Rivers.

Predictably, once activities and uses of the different watercourses started to become more specialized, so did their governing treaties and the organs created by them. Hence, the treaty celebrated between Argentina and Paraguay to conduct studies on the technical and economic feasibility of multiple uses particularly in their stretch of the Paraná River, also created its own institutional structure; the Mixed Commission that we have already referred to above as the COMIP.

The COMIP has a particular feature, in that it was recognized in the treaty that created it as an international organization, with sufficient legal capacity to acquire rights, be bound by obligations and carry out any and all acts necessary in order to achieve its goals.[30] It is the only mixed commission in the Paraná Basin that has been recognized as an autonomous international organization. It has actual competence to act independently from the directives of its member States, governing over activities including power generation, navigation, fishing, industrial and agricultural use of the waters, recreational uses and the protection and preservation of the portion of the Paraná that is subject to its rule.

In fact, based on further agreements between Paraguay and Argentina, the COMIP has expanded its competencies with time. For instance, the 1979 Tripartite Agreement signed by Argentina, Paraguay and Brazil included the power to exchange hydrological data with the ITAIPU Binational Entity, and to verify that the parameters agreed to in this Agreement are being observed at the Itaipu plant, upstream. Furthermore, in 1989 the COMIP was empowered to conduct studies on water quality and fishing resources in the Paraná River.[31] In 1992,

Argentina and Paraguay decided that there should be only one commission in charge of the entire management of their stretch of the Paraná River. The COMIP was appointed as the Administrating Agency of the River, further extending its powers to include oversight of navigation and environmental control.[32]

Also, in 1992, the Parties tasked the COMIP with the drafting of a Statute for the Paraná River, in order to include all uses of the river in one single instrument. However, no statute has been adopted to date. In 1996, a new Agreement was signed between Argentina and Paraguay for the preservation and development of fish resources throughout their transboundary sections of the Paraná River and Paraguay River.[33] This Agreement created a Coordinating Committee, but it names the COMIP as its Permanent Secretariat.[34]

Notably, the COMIP has all but one area of competence; the generation of hydropower at the Yaciretá dam. This was ceded to the Yaciretá Binational Entity (EBY, by its Spanish acronym). The same occurred at its upstream sister plant, the Itaipu dam, which is governed exclusively by the Itaipu Binational Entity (ITAIPU). Each of these entities has the power to conduct all technical activities for the exploitation of hydropower in their respective plants, as well as to manage the distribution and sale of its product to the power management entities of each of the States involved.

One singularity is that both entities are in fact a partnership between state power management entities. In the case of ITAIPU, the partners are the Administración Nacional de Electricidad (ANDE) of Paraguay, and the Centrais Elétricas Brasileiras S.A. (Eletrobrás) of Brazil, and each partner has an equal share in participation in ITAIPU's capital.[35] ITAIPU has headquarters in the respective capital cities of Brazil and Paraguay, and is managed by an Administrative Council and an Executive Board composed by the same number of nationals of each State, that is, 12 Councilmen, six designated by Paraguay and six by Brazil. ITAIPU is in charge of the management of the plant and the administration and government of the treaty, its annexes, additional protocols and any other documents derived therefrom.

The entity that manages Yaciretá, EBY, has the same legal nature and powers as ITAIPU. It is a partnership integrated by A. y E. and ANDE,[36] with equal participation in the entity's capital. EBY has headquarters in the Paraguayan and Argentinian capital cities, and it is also governed by an Administrative Council and an Executive Board, each composed of an equal number of nationals from each State.[37]

The Guarani Aquifer Agreement (GAA) contemplates the creation of another Commission comprised of its four State Parties (OAS, 2009). The Commission shall coordinate the cooperation among such Parties for complying with the principles and objective of the Agreement.[38] To date, this Commission has not been created, since the treaty itself has not yet entered into force.

The Paraná River is but one of the three main watercourses in the Rio de la Plata basin, and the fact that the same State Parties are involved inevitably means that there will be interaction and coordination with institutions created by treaties that do not directly govern a part of the Paraná. In this connection, we must name the Mixed Technical and Executive Commission for Dredging the Paraguay River, which has a mandate to maintain cooperation with the CIH, since the Paraguay River is a component of the Waterway. In this case, Paraguay and Argentina have given the Mixed Commission powers to ensure that the Paraguay River is kept in a condition that is optimal for navigation. Obviously, these activities directly influence the operation of the Waterway, and so the meetings of both organs are usually held in the same city, on the same dates.[39]

There are treaties affecting tributaries of the Paraguay River that have resulted in the creation of institutional structures. The Bermejo River Treaty has two organs; the Binational

Commission for the Development of the High Basin of the Bermejo River (COBINABE) and the Regional Commission of the Bermejo River (COREBE). Both of these organs have members from Argentina and Bolivia. Although there is no evidence of cooperative activities between these two Commissions and the rest of the organs that have activities in the Rio de la Plata Basin or the Paraná Basin, it seems inevitable that this will change in the near future. In fact, Argentina and Bolivia decided in 2017 to resume their projects for the construction of hydropower plants in the Bermejo River, which will ultimately affect the conditions downstream that include the Paraguay River and the Paraná River, as well as the Waterway.[40]

Mixed commissions have also been created for the Apa River and the Pilcomayo River. For the Apa River, Paraguay and Brazil created a Mixed Commission for the Sustainable Development and Integrated Management of the Apa River Basin.[41] It has headquarters in Bela Vista in the state of Mato Grosso do Sul, Brazil and Bella Vista, Paraguay. One particular feature of this Apa River treaty, and the Commission created by it, is the fact that the regime welcomes the notion of Integrated Water Resources Management (IWRM), which is a concept that seeks to include human rights and environmental practices within a comprehensive approach to manage water resources (Bernex et al., 2017). We have not found any evidence of cooperation between this Mixed Commission and others within the Rio de la Plata Basin, or the Paraná Basin.

The Pilcomayo River Treaty created a Tripartite Commission formed by two delegates from each of the three State Parties: one with a political function and the other with technical functions. The Tripartite Commission has full power to oversee the execution of all projects that the parties may wish to implement in the River.

The Commission will conduct environmental studies and develop a project for the equitable and reasonable management of the natural resources of the Pilcomayo.[42] It is also worth mentioning that two more institutions are inevitably thrown into the mix, since they were created by Uruguay and Argentina to regulate activities in the Uruguay River and the Rio de la Plata: Respectively, the Administrative Commission for the Uruguay River (CARU, by its Spanish acronym), and the Administrative Commission for the River Plate (CARP). These two commissions have roughly similar purposes and functions as the rest, and in practice, interact with the other institutions, particularly in relations to what pertains to the Waterway.

Examples of institutional cooperation within the basin

Having reviewed the vast number of institutions that oversee activities in and around the Paraná River Basin, this section will examine a few examples of how these institutions interconnect and cooperate.

Cooperation in activities associated with the generation of hydropower

Three treaties were signed to oversee hydropower uses and the Commissions created for the Paraná River.[43] In fact, the ITAPU Binational Entity,[44] the EBY,[45] and the COMIP,[46] all operate simultaneously in controlling and managing the activities associated with the study, construction, operation and management of the product of their respective dams.

These three organs support each other in activities such as monitoring the water quality of the Paraná River. For instance, in designing the Water Quality Monitoring Network for the process of filling the reservoir, EBY received assistance from COMIP.[47] In addition, all three Commissions cooperate in designing campaigns and software, such as the GNoma,[48] to improve the study of water quality and measurements.

Cooperation in the Paraná-Paraguay waterway

The Waterway Agreement created the CIH as a political organ and the Commission of the Agreement as its technical organ, which oversees the maintenance of the entire Waterway. At the same time, the Paraguay River Treaty created the Technical Executive Mixed Commission, which is in charge of regulating, channeling, dredging, signaling and maintaining the Paraguay River, which is the upstream half of the Waterway. Simply viewed, the operations and competencies of these two institutions might overlap and cause confusion. However, for instance, the Waterway Agreement's Regulation No. 2 about signaling has not been internalized by Argentina or Paraguay,[49] so, in practice, both Commissions act independently from one another, but complementing each other's functions.[50]

Cooperation through FONPLATA

The Financial Fund for the Development of the Rio de la Plata Basin (best known as FONPLATA) is a multilateral development bank with international legal status. Its purpose is to support the integration and harmonious and inclusive development of the member States for their better integration within the region and the global market, by funding studies, projects, programs, assistance and technical advice.[51]

Since its creation in 1974, FONPLATA has fulfilled its purpose by financing the execution of studies, projects, programs and works in the Rio de la Plata Basin. It has prioritized projects aimed at building physical infrastructure to complement the existing regional systems; social investment in education, health, basic infrastructure and water distribution; agriculture, livestock and industrial production; promotion of exports; environmental projects; protection and preservation of natural resources; treatment of polluted water; etc. FONPLATA grants provide funds directly to member States or to its agencies and organs, but this fund also participates in financing regional projects.

Cooperation between CIC and COMIP

On April 12, 2019, the Management Director of the COMIP attended a meeting at CIC Headquarters in order to coordinate efforts between both Commissions to promote the harmonious development and physical integration of the Rio de la Plata Basin and avoid redundancy in their actions. It was reported that both institutions identified critical common issues, as well as opportunities for joint development.

The Secretary General of the CIC emphasized the need for a Basin Observatory, integrated by technical professionals of each area of interest, and a digital library available to both institutions, in order to centralize information and make its use more efficient.

The two Commissions agreed to provide each other with and exchange publications and technical studies, as well as possibly conduct certain technical studies jointly. The two organs will also celebrate operational and legal agreements in order to strengthen and coordinate actions and improve regional cooperation.[52]

Cooperation challenges

With the one exception of the Itaipu Treaty, which is exclusively directed at hydropower generation, every other international agreement existing in the Paraná Basin and its surroundings includes a multiplicity of uses. While this ensures that virtually no human activity goes unregulated in the Paraná basin, it presents a number of issues when two or more legal norms collide, and there is no clear method of resolution for such conflict between norms.

A first possible conflict may arise when some of the bilateral and multilateral agreements examined in this chapter have also created their own system for conflict resolution and others simply have left that area silent. Therefore, the first difficulty in determining a solution for a conflict of law scenario is precisely which organ, if any do exist, would be competent to give a solution, and which procedure would be applicable. Secondly, what would be the applicable law? If there is a conflict of norms between the Waterway Agreement and the Paraguay Agreement, which treaty would have preferential application? This section studies some of the potential areas of conflict among the treaties of the system, which might present challenges to the efforts of cooperation in the Paraná basin.

The Waterway Agreement and the Paraguay River agreement

As described in the previous sections, the Waterway Agreement provides an ample set of rules for the operation of the Paraná-Paraguay Waterway, with all five states as parties to the agreement. On the other hand, the Paraguay River Agreement for regularization, channeling, dredging, signage and maintenance of the Paraguay River covers only the stretch of the river that constitutes the boundary between Argentina and Paraguay, and, of course, only these two States are parties to this treaty.

In practice, the activities conducted by Argentina and Paraguay within the scope of the Paraguay River Agreement have an important effect on the activities of the Waterway, which affect the entire area, north to south, and all five States, and vice versa. The Mixed Technical and Executive Commission of the Paraguay River Agreement executes the obligations under the Paraguay River Agreement, while the preservation of conditions of the rest of the Paraguay and Paraná rivers falls to the CIH.

It is crucial that both Commissions read and interpret their governing treaties in consonance with that of the other. Otherwise, there is a risk of costly redundant actions or, even worse, no action at all. As a solution, the Waterway Agreement's Fifth Protocol contains rules specifically about the activities contemplated in the Paraguay River Agreement, which calls for coordination among State Parties in case of conflict of norms.[53]

The Waterway Agreement and the Treaty of the Uruguay River

The Treaty of the Uruguay River is outside the scope of the Paraná Basin. However, geographically it overlaps jurisdictions with the Waterway Agreement, because the Waterway actually ends in the Port of Nueva Palmira, which is not a port within the Paraná River, but belongs to the Uruguay River. The same port marks the terminus of the Paraná-Paraguay Waterway, and the beginning of the Uruguay River Waterway, which is completely independent from the former. This explains why Uruguay is a party to the Waterway Agreement, but it also poses a risk for overlapping competencies between the CIH and the CARU, respectively the administrating organs of the Waterway and the Uruguay River.

The Uruguay Statute, which regulates the uses of the Uruguay River, includes provisions that deal with navigation, such as signaling and dredging. These provisions could collide with other rules on the same subject matters contained in the Waterway Agreement. For instance, Article 5 of the Uruguay River Statute provides that the CARU has the power to assign to the parties of that treaty their share of obligations regarding dredging, signaling and other preservation activities in the main navigational channel of the river. This, of course, includes the port of Nueva Palmira. If the assignments made by CARU should collide with norms of the Waterway Agreement, the only way to resolve this conflict of law is to resort to the general rules of International Law applicable to conflicts of law. In this connection, on the one hand, the Uruguay Statute was signed in 1976, while the Waterway Agreement came into force in 1995, thus the posterior norm prevails. The Waterway Agreement also includes regulations that are more specific than the general norms of the Statute, so again the Waterway Agreement would prevail. On the other hand, if the Statute were deemed as a more specific norm due to the number of parties, or its more limited territorial application, then the Statute would prevail. A practical consequence of this potential conflict of norms would be the delay in performance of activities that are necessary to ensure optimal navigation conditions both in the Paraná-Paraguay Waterway and the Uruguay River Waterway. Another, more palpable, consequence is the delay in the construction of grain terminals at the Port of Nueva Palmira, which has been awaiting authorization from the CARU for several years.

The Paraná River treaty and the Yaciretá treaty

Another interesting case where conflict might occur is where the Paraná River Treaty and the Yaciretá Treaty converge, since they both cover the same territory. As a reminder, the first is overseen by the COMIP, while Yaciretá is governed by the EBY. The source for conflict here is limited to hydropower generation. However, the parties to both treaties aptly excluded that subject matter from the competencies of the COMIP. The relationship between these two Commissions is governed by a Framework Agreement of Cooperation, signed by both institutions in 2001.

The purpose of this instrument is to ensure '*... joint participation of both organs in the exchange of technical information, training of human resources, and implementation of studies and actions in accordance with the inherent functions of each organ*'.[54] Faced with a possible conflict or challenge, it would seem that the COMIP, an apparently lesser institution compared to the CIC or the CIH, would have the lead in moving forward and achieving more organized and formal means of cooperation with its sister and cousin agencies.

The apparent fragmentation of the legal framework and the institutional organization of the entire system

The practice of the States involved in the Paraná Basin and the Rio de la Plata Basin to date seems to indicate that the tendency to fragment the law and compartmentalize activities is here to stay. In fact, in a study conducted by the OAS about a potential framework for the sustainable management of the River Plate Basin, it is stated that: '*except in exceptional cases, there are no mechanisms for communication among* [the organ of the system] *or with the CIC*'.[55] The study cites the creation of the CIH supposedly within the framework of the River Plate Basin Treaty, which in practice acts independently and without any coordination with the CIC.

In order to confront this apparent weakness of the system, the Meeting of Foreign Ministers of the Rio de la Plata Basin agreed to amend the Statute of the CIC in 2001. The current version of the Statute calls for the coordination of projects led independently by other organs in the area with the CIC. A deeper study of the practice of the institutions involved in the Paraná River Basin – particularly the COMIP, the EBY and ITAPU – allows for the perception that there is a true spirit of cooperation in the area.

Prospects for improved relations

Despite the just mentioned areas of cooperation, there is certainly room for improvement in the management of the water resources in the region. This section discusses some of the prospects for improved relations in the area.

The conflict between limited sovereignty over shared water resources versus permanent sovereignty over natural resources

One of the conflicts that requires prompt attention in the Paraná River Basin is the situation of Paraguay versus Brazil and Argentina in the Treaties of Itaipu and Yaciretá, respectively. As mentioned above, for years Paraguay has been trying unsuccessfully to vindicate its right to fully enjoy its share of the power generated by the Itaipu and the Yaciretá hydropower plants.

Both treaties provide that any surplus power can only be sold, or in the case of Yaciretá, ceded to the other party at prices well below market value.[56] Paraguay insists that the sale of its surplus energy at competitive prices would greatly aid its economic development.[57] The Paraguayan claim views that, in the case of Itaipu, this vindication would actually be a return to the initial agreement between Paraguay and Brazil, which gave the latter a right of preference in the purchase of the surplus but did not establish a prohibition to sell to third parties. Although talks are ongoing, Annex C of the Itaipu Treaty will only be open to amendments in 2023. Paraguay argues that Yaciretá reproduced exactly the same conditions as Itaipu, and this leads to Argentina obtaining massive benefits from the Paraguayan surplus.[58] The term for amending Annex C of the Yaciretá Treaty opened in 2013, and the parties have been in negotiations ever since, without any result.

Clearly, the situation between Paraguay and its counterparts raises the questions of whether, in these cases, there is a true application of the principle of equitable and reasonable use of a water resource. In addition to that, this presents an undue unbalance between the obligation to share sovereignty over riparian resources, and the right to permanent sovereignty over natural resources, which is, in essence, what Paraguay is claiming. The current state of International Law seems to support the notion that electricity produced via hydropower is a natural resource secondary to the water resource, because it is a by-product. Under such a concept, Paraguay would indeed have a claim of full sovereignty over its 50% share of the energy produced at Itaipu and Yaciretá (Schrijver, 1997; Szabo, 2016). International arbitration awards have concluded that sovereignty over natural resources grants the States exclusive competence to organize their economic structures without foreign intervention.[59] Could Paraguay have a right to use its surplus energy as it sees fit? Currently, the treaties involved do not offer a means of resolution other than diplomacy, but if Paraguay cannot get satisfaction, it might seek sufficient legal grounds to be heard by an international tribunal in the future.

Non-ratification of the existing international instruments

None of the states of the Rio de la Plata Basin or the Paraná River Basin have ratified the 1997 Convention on the Law of Non-Navigational Uses of International Watercourses (the UN Watercourses Convention),[60] or the 1992 United Nations Economic Commission for Europe Convention on the Protection and Use of Transboundary Watercourses and International Lakes, hereinafter referred to as the UNECE Water Convention.

In 2016, the UNECE Water Convention was opened for signature by non-European States.[61] It is difficult to pinpoint the reasons why the countries sharing the Paraná basin have not ratified and do not show interest towards these international legal instruments, especially considering that the regulation of the use of water resources in the region predates these international conventions. However, this may be exactly the reason for this lack of enthusiasm for participating in the international conventions. It could be simply a matter of believing that the existing framework already works for the states governing the Rio de la Plata Basin and the Paraná basin. Alternatively, the problem may be that the scope of the UN Watercourses Convention is broader than the regional framework and includes all human acts that may be carried out in the watercourses. A detailed review of the treaties that regulate the uses of the watercourses of the Paraná basin, and more broadly the Rio de la Plata Basin, may well confirm that this is indeed the reason for the State Parties to avoid ratifying a Convention, because of its broader scope.

Sustainability of the Paraná River Basin framework in its current state

Throughout this chapter, several examples have been presented that demonstrate the cooperation and coexistence that are the result of Paraná River Basin treaties. When drafting the legal instruments, the parties were careful not to allow competencies and powers to overlap. The institutions that coexist, including ITAIPU, EBY and COMIP, are examples of efforts in cooperation, which have already begun to positively influence other organs of the Rio de la Plata Basin. Despite opinions that consider the legal framework in the region to be a 'tangled mess',[62] our review reveals the harmonious existence of over 20 different organs, deriving from nearly 15 multilateral and bilateral treaties, which in over 35 years have not resulted in conflict or have been a cause for failure.

There is certainly room for further efforts in cooperation, which seem to be occurring on their own, which denotes a spirit of coordination among the parties that is not apparent at first glance. There does seem to be an order to this regional chaos, in that somehow the State Parties have managed to create a working, effective legal framework. Some improvement is due regarding the fact that, while the CIC has managed to coordinate the actions of the rest of the states, its work is mainly focused on national and local entities, instead of the 14 Technical and Administrative Commissions that were created by these States to manage activities in each area of the basin. One of the challenges of the CIC will be to plan for sustainable management of water resources within the context of climate change (CIC Plata, 2017). In our view, the entire region would greatly benefit from continuing the work of the COMIP to promote the exchanging and sharing of information, training and other activities.

Conclusions

This chapter has illustrated the very complex legal picture of governance within the Paraná River Basin. The basin is governed by a plethora of legal instruments, some bilateral and some multinational, including the framework Rio de la Plata River treaty, energy specific projects such as ITAIPU, the yet to come into force Guarani Aquifer Agreement and treaties related to other tributary rivers such as the Paraguay River, and with the neighboring Uruguay River basin.

Thus, the complexities of the legal system confronting any stakeholder are bewildering. This chapter serves as a compass to navigate through such treacherous waters. What stems from the analysis is the existence of a fragmented legal system, but one that does seem to work. The plethora of commissions and institutions created by the different legal instruments could potentially lead to dangerous conflicts of norms, which could have harmful effects on the proper management of the Paraná. However, such conflicts have not yet surfaced and the signatory countries have been able to cope with the management of the Paraná quite successfully up to the present. This does not mean that there are not any problems. The imbalance that Paraguay suffers when it comes to its energy relations with Brazil and Argentina is probably the most striking of these ongoing challenges that will require resolution.

Acknowledgements

We thank Dieison André Moi for providing the photograph of the Itaipu dam, by authorization from Itaipú Binacional.

Notes

1 Rio de la Plata Basin Treaty, Preamble. Note: quotes from treaties are originally in Spanish, and have been translated by the authors.
2 See Article VI, which allows the State Parties to celebrate specific or partial agreements, whether bilateral or multilateral, aimed to achieve the general purposes of development of the Basin.
3 See operative clauses 1–7 of Resolution 25.
4 Resolution No. 210 of the Meeting of Foreign Affairs Ministers.
5 Declaración de La Paz, August 6, 1989.
6 Montevideo Treaty of 1980, and Articles 2 and 10 of Resolution No. 2 of ALADI'S Council of Ministers.
7 Available at: www.aladi.org/nsfaladi/textacdos.nsf/6de700ceb7fe34a3032567f200721748/8213df518a1151db0325682500718c7f?OpenDocument.
8 Available at: www.comip.org.ar/wp-content/uploads/2017/05/Convenio-para-el-estudio-del-aprovechamiento-de-los-recursos-del-ri%CC%81o-parana%CC%81..pdf.
9 Article 1.
10 Availableat:www.gwp.org/globalassets/global/gwp-sam_files/programas/dai/manual-derecho-de-aguas-internacionales.pdf, Page 46.
11 https://www.itaipu.gov.py/es/sala-de-prensa/faq.
12 Available at: www.itaipu.gov.py/es.
13 Artículo V of the Treaty, the Contracting Parties grant the ITAIPU Binational Entity a concession to carry out the exploitation of the Parana River's resources for the generation of power. See: www.itaipu.gov.py/es/institucional/documentos-oficiales.
14 See Article in Color ABC, available at: https://www.abc.com.py/edicion-impresa/economia/el-consumo-de-electricidad-alcanzo-un-nuevo-pico-1780791.html.
15 Source: Multiple Declarations and news releases throughout the years. The following is a link to the declarations of the former Director of the Paraguayan side of the dam, Carlos Mateo Balmelli: www.abc.com.py/edicion-impresa/economia/se-buscara-precio-justo-y-venta-de-energia-a-terceros-paises-1083086.html.
16 Available at:https://www.mre.gov.py/v2/Adjuntos/Normativas/TRATADOYACYRETA.pdf.
17 This was accomplished by an Exchange of Diplomatic Notes between both State Parties.
18 Article XXII.
19 Source:www.abc.com.py/edicion-impresa/economia/el-50-de-la-energia-de-yacyreta-es-de-paraguay-pero-recibio-57-1431093.html.
20 http://www.nanduti.com.py/2017/02/27/piden-revisar-la-deuda-de-yacyreta-antes-de-iniciar-obras-de-ana-cua/.
21 Source: www.elterritorio.com.ar/la-nacion-buscaria-regular-la-explotacion-del-acuifero-guarani-9507122764930137-et.
22 Available at: www.internationalwaterlaw.org/documents/regionaldocs/Guarani_Aquifer_Agreement-Spanish.pdf.

23 Articles 3 to 12.
24 Available at: www.internationalwaterlaw.org/documents/regionaldocs/Guarani_Aquifer_Agreement-Spanish.pdf.
25 As expressed in the introduction to the River Plate Basin Treaty, published by the CIC. Available at: http://cicplata.org/wp-content/uploads/2016/12/tratado-de-la-cuenca-del-plata.pdf.
26 The position of the then Brazilian Foreign Minister was that the CIC should be an agency for coordination of the positions of all member States, and should act under their direction, without sacrificing any portion of their sovereignty and interests.
27 River Plate Basin Treaty, Article II.
28 Meeting of Public Works and Transportation Ministers, Santiago de Chile Act, September, 1989.
29 Resolution 238 of the 20th Meeting of the States of the River Plate Basin Treaty.
30 Article 4.
31 Agreement by Exchange of Diplomatic Notes between Paraguay and Argentina about Fish and Water Quality, 1989. Available at: www.comip.org.ar/wp-content/uploads/2017/05/1989-ACUERDO_FAUNA.pdf.
32 Agreement by Exchange of Diplomatic Notes between Paraguay and Argentina naming the COMIP as the Administrating Agency of the Parana River, 1992. Available at: www.comip.org.ar/wp-content/uploads/2017/05/1992-ACUERDO-Comisio%CC%81n-administradora.pdf.
33 Signed October 25, 1996. Available at: www.bacn.gov.py/leyes-paraguayas/2483/aprueba-el-convenio-sobre-conservacion-y-desarrollo-de-los-recursos-icticos-en-los-tramos-limitrofes-de-los-riosparana-y-paraguay.
34 Article 12.
35 Itaipu Treaty, Article III, paragraph 2.
36 A. y E. is the Argentinian State Waterworks Company, and ANDE is the Paraguayan National Electric Power Agency. Yaciretá Treaty, Article II(e) and II(f).
37 Article IV.
38 Guarani Aquifer Agreement, Article 15.
39 See, for instance, the Resolution 542 of the Paraguayan Ministry of Foreign Affairs, of June 1, 2017, authorizing its representative to travel to Argentina to participate in both meetings.
40 Source: www.enernews.com/nota/304211/represas-del-rio-bermejo-activacion-y-polemicas
41 Article II of its Statute defines it as the '*binational organ responsible for the execution of the Cooperation Agreement between the Republic of Paraguay and the Federative Republic of Brazil for the Sustainable Development and the Integrated Management of the Apa River Basin*'.
42 Pilcomayo Treaty, Article IV.
43 The dam projects under the Itaipu and Yacyretá Treaties, and the Parana Agreement are all in the Parana River.
44 Visit the ITAIPU websited for more information about its activities at: www.itaipu.gov.br/es.
45 See current activities at: www.eby.org.ar/y www.eby.gov.py/.
46 The COMIP's activities can be seen at: https://www.comip.org.ar/.
47 See full text at: www.eby.org.ar/index.php/medio-ambiente/calidad-de-agua.
48 See: www.comip.org.ar/calidad-de-agua/#1550156788143-e3fb8657-d1be
49 See: http://www2.aladi.org/nsfaladi/textacdos.nsf/vArt14webR?OpenView&Start=1&Count=100&-Expand=4#4.
50 On June 7, 2017, the Mixed Technical and Executive Commission for the Paraguay River held its annual meeting in Buenos Aires, while the CIH meeting had been held just days earlier in the same city. While each Commission acts independently, the same representatives attend both meetings, so the member States ensure that there is no actual overlap, but rather the activities of one Commission complement those of the other. See: https://www.icndiario.com/2017/09/paraguay-y-argentina-acuerdan-iniciar-tareas-de-dragado-del-rio-paraguay/.
51 Articles of Agreement, originally signed at the VI Ordinary Meeting of Ministers of Foreign Affairs of the RiverPlate Basin States, on June 12, 1974 (Buenos Aires, Argentina), Articles 1 and 2.
52 See full report at: www.comip.org.ar/blog/category/noticias-comip/.
53 Additional Protocol to the Paraná-Paraguay Waterway Agreement about Navigation and Safety, Article 36.
54 Yacyretá-COMIP Framework Agreement of Cooperation, Article 1. Available at:www.comip.org.ar/wp-content/uploads/2017/11/002CONV-EBY-COMIP-2001-RENOVAC-AUTOMAT.pdf.
55 Framework for the Sustainable Management of the Water Resources in the River Plate Basin, Legal and Institutional Framework. Available at: www.oas.org/dsd/plata/legalf.htm.

56 Office of the President of the Paraguayan Delegation at Parlamento de Mercosur, Declaration of May, 2012. Available at: www.parlamentomercosur.org/innovaportal/v/6296/1/parlasur/paraguay-recla-ma-precio-justo-por-energia-de-itaipu.html.

57 ABC Paraguay, El Primer Reclamo Paraguayo es la Plena Disponibilidad de su Energía, March 20, 2017, Available at: www.abc.com.py/edicion-impresa/economia/el-primer-reclamo-paraguayo-es-la-ple-na-disponibilidad-de-su-energia-1575459.html.

58 Office of the President of the Paraguayan Delegation at Parlamento de Mercosur, ARLASUR, Declaration of May 2012.

59 Texaco Vs. Libyan Arab Republic, 17 ILM (1978), Paragraph 59.

60 See ratification status at: https://treaties.un.org/pages/ViewDetails.aspx?src=TREATY&mtdsg_no=XXVII-12&chapter=27&clang=_en.

61 See ratification status at: https://treaties.un.org/Pages/ViewDetails.aspx?src=TREATY&mtdsg_no=XXVII-5&chapter=27&clang=_en.

62 Available at: www.lanacion.com.ar/1864565-una-marana-legal, last visited February 4, 2018.

References

Aguilar Rojas, G., Iza, A. 2009. *Gobernanza de Aguas Compartidas: Aspectos Jurídicos e Institucionales, International Union for Conservation of Nature (IUCN).* IUCN, Gland, Switzerland.

Amore, L. 2011. The Guarani aquifer: From knowledge to water management. *Water Resources Development* 27:463–476.

Bernex, N., Colom de Morán, E., Garcia-Pachón, M.P., Hantke-Domas, M., López, A., Pinto, M., Sanchez, J.C. 2017. El Derecho Internacional de las Aguas en América Latina, Manual de Capacitación, Asociación Mundial para el Agua. Retrieved from: www.gwp.org/globalassets/global/gwp-sam_files/programas/dai/manual-derecho-de-aguas-internacionales.pdf

CIC Plata. 2017. *The framework program for the sustainable management of the La Plata Basin's water resources, with respect to the effects of climate variability and change,* Strategic Action Program for the La Plata Basin, Organization of American States, Washington, D.C., USA.

Gilman, P., Pochat, V., Dinar, A. 2008. Whither La Plata? Assessing the state of transboundary water resource cooperation in the basin. *Natural Resources Forum* 32:203–214.

Laborde, L.C. 2008. The Rio de la Plata River basin: The path towards basin institutions. In: Varis, O., Tortajada, C., Biswas, A.K. (Eds.), *Management of Transboundary Rivers and Lakes,* Springer-Verlag, Berlin/Heidelberg, pp. 269–289.

McCaffrey, S. 2007. *The Law of International Watercourses,* 2nd ed., Oxford: Oxford University Press.

Newman, S.J. 2010. Capitalist development as White Elephant: A case study of Argentina's Yacyretá hydro-electric dam. *Eleven: The Undergraduate Journal of Sociology* 1:54–83.

OAS. 2009. *Guarani aquifer: Strategic action program. Acuífero Guaraní: Programa Estratégico de Acción,* Bilingual Edition, Organization of American States (OAS), Washington, D.C., USA.

Querol, M. 2002. Estudio sobre los Convenios y Acuerdos de Cooperación entre los Países de América Latina y el Caribe, en relación con Sistemas Hídricos y Cuerpos de Agua Transfronterizos, CEPAL, Serie Recursos Naturales e Infraestructura. Retrieved from: http://archivo.cepal.org/pdfs/Waterguide/lcl2002s.pdf

Rama Montaldo, M. 1972. Cuencas internacionales y cuenca del Plata. *Revista Uruguaya de Derecho Internacional* 1:115–136.

Schrijver, N. 1997. *Sovereignty Over Natural Resources: Balancing Rights and Duties.* Cambridge: Cambridge University Press.

Sindico, F. 2011. The Guarani Aquifer system and the International Law of Transboundary Aquifers. *International Community Law Review* 13:255–272.

Sindico, F., Hirata, R., Manganelli, A. 2018. The Guarani aquifer system: From a beacon of hope to a question mark in the governance of transboundary aquifers. *Journal of Hydrology Regional Studies* 20:49–59.

Sorgine, G. 2012. O caso Itaipu: Estratégias e legitimação nas relações Brasil-Paraguai. *Conjuntura Internacional* 9:41–49.ç.

Szabo, (2016) The EU Member States' Right to Electricity Mix. DOI:10.5817/MUJTL2016.

Villar, P.C., Ribeiro, W.C., Sant'Anna, F.M. 2018. Transboundary governance in the La Plata River basin: Status and prospects. *Water International* 43:978–995.

9

Meeting the Sustainable Development Goals for water and sanitation

Chris D. Metcalfe and José Esteban Castro

Key messages

- Although the countries that share the Paraná River Basin, namely Brazil, Argentina and Paraguay, have officially stated that they are on track to meet the targets of SDG for universal access to affordable safe drinking water and sanitation, and eliminating open defecation, the available evidence suggests that these countries will fail to achieve these targets.

- Despite reporting high levels of service coverage in urban areas, official figures must be interpreted with caution. Reported coverage levels often misrepresent the situation affecting important segments of the urban population, for example those living in shanty towns and favelas. Moreover, the quality of water delivered for human consumption in urban areas too often fails to meet minimum standards, and this is not adequately reflected in official reports. Consequently, people resort to expensive and mostly unregulated bottled water to satisfy their needs.
- The countries within the basin face significant challenges in addressing SDG 6 to provide universal access to facilities for sanitation and hygiene, particularly among the poorer segments of the population that live in rural areas or in urban slums. Afro-American and Indigenous communities are particularly affected.
- All three countries will have to make significant investments to treat domestic and industrial wastewater to meet the targets of SDG 6.3.
- Argentina and Paraguay need to develop programs to monitor water quality and report on these data to the UN-Water Integrated Monitoring Initiative for SDG 6 (IMI).
- Implementation of programs for integrated water resource management appears to be at or below the global average in the countries that share the Paraná River Basin (i.e. <50%), but there are significant barriers to increasing the participation in these types of water management programs.
- The Forum of the Countries of Latin America and the Caribbean on Sustainable Development convened under the auspices of the Economic Commission for Latin America and the Caribbean (ECLAC) may be an effective mechanism for monitoring and reviewing the implementation of the 2030 Agenda for Sustainable Development among the Member States, including reporting progress in Brazil, Argentina and Paraguay in meeting the objectives of SDG 6.

Introduction

World leaders adopted the 2030 Agenda for Sustainable Development, which set 17 goals for improving the lives of the global population and for protecting the environment. Sustainable Development Goal 6 (i.e. SDG 6) focuses on access to affordable clean water, sanitation and hygiene for all. However, water serves as a foundation for many of the other development goals (UN-Water, 2016), such as SDG 3 (Good Health and Wellbeing), SDG 11 (Sustainable Cities and Communities), SDG 13 (Combat Climate Change), SDG 14 (Life Below Water) and SDG 15 (Life on Land). As summarized in Figure 9.1, there are eight targets identified within SDG 6. These include Target 6.1 that aims to provide access to safe drinking water for 100% of the population in each country, and Target 6.2 that aims to provide 100% access to facilities for sanitation and hygiene. All other objectives within SDG 6 are 'aspirational' targets, meaning that individual participating countries can set their own targets and report progress relative to those objectives. Targets 6.a and 6.b are organizational objectives, as opposed to other operational SDG 6 objectives for the participating countries.

However, according to the Synthesis Report on Water and Sanitation published in July 2018 (UN-Water, 2018), there will be significant challenges for many countries to achieve the targets set for SDG 6. Three years into the SDG era, this report confirms that the world at large is lagging behind in achieving the SDG 6 targets. Less than half of the UN Member

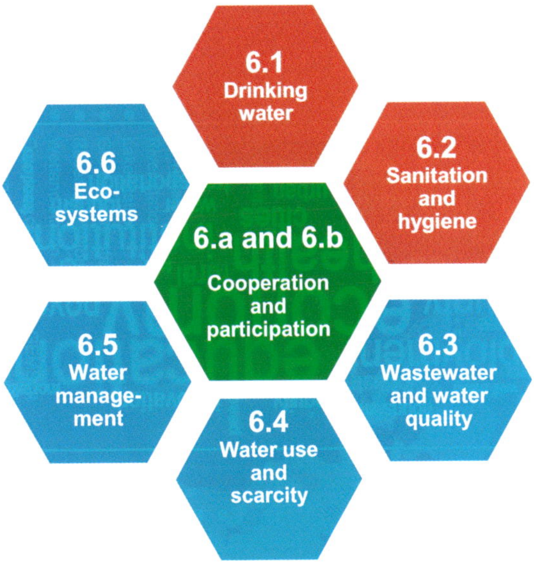

Figure 9.1 Targets included within Sustainable Development Goal 6. Hexagons in red indicate target data collected by JMP. The hexagon in green indicates target data collected by GLAAS. All other target data represented by blue hexagons are collected by GEMI.

States have data available on progress made towards SDG 6 targets, and there is a lack of tools available for countries to effectively track the indicators for SDG 6 (UN-Water, 2018). In addition, while low- and middle-income countries may focus on advances in achieving some of the SDG objectives, such as reductions in poverty (SDG 1), improvements to health (SDG 3) and increased economic growth (SDG 8), advances in SDG 6 may not be a high national priority in the competition for limited financial resources.

Investments in infrastructure for the treatment of water and wastewater and the development of policies and practices that reduce point-source pollution are generally a low national priority relative to other investments. In a survey of 70 low-, middle- and transitional-income countries, only 19% and 9% had sufficient financial resources to meet water quality targets in urban and rural areas, respectively (GLAAS, 2017). Even in countries such as the new member states of the European Union, where membership in the Union requires compliance with the Water Framework Directive to meet requirements for 'good status' of their water resources, progress has been slow in improving water quality (Teodosiu et al., 2015). Wastewater management has historically been poorly funded because it has been considered less visible, with lower public recognition and less political support (Winpenny et al., 2016). A recent study with a focus on countries in Latin America showed that policies that attract clean and energy efficient industries have the potential to improve environmental quality while also enhancing economic growth (Sapkota and Bastola, 2017). However, many countries do not have the luxury of attracting only clean industries, and instead have a high proportion of polluting industries. The '*pollution haven*' hypothesis holds that polluting industries will re-locate from countries with more stringent environmental regulations to countries that have lax regulations and/or ineffective enforcement (Ederington et al., 2005; Tang, 2015).

Although Latin American countries have a very diversified approach to priorities for investment, environmental sustainability tends to be a low priority. In most cases, national goals tend to focus on economic growth, but in some cases extend to improving education, reducing poverty and extending access to essential public services. These latter goals, which are complementary and mutually reinforcing with SDG 6, have had a higher priority in the last two decades. This was the case, for example, in Argentina and Brazil during the period 2008–2015. Nevertheless, the lower priority for goals focused on environmental sustainability is illustrated by the lexicometric analysis conducted in 2018 by the Economic Commission for Latin America and the Caribbean (ECLAC) on the key concepts present in national development plans (Figure 9.2). However, despite the focus within Latin American countries on improvements to the economy, health and education, and in reducing poverty, there is evidence that SDGs related to the environment are slowly becoming integrated into long-term national goals, at least where these objectives converge with other national priorities. In Brazil, access to sanitation was included in a national agenda for economic activities (Rocha and Weiss, 2019). In Brazil and Argentina, advances in SDG 6 are considered integral to making progress towards the health-related SDGs (GBD 2016). According to ECLAC (2018), among countries from Latin America and the Caribbean that were polled concerning their priorities under SDG 6, the target related to access to sanitation and hygiene (i.e. SDG 6.2) was prioritized in all cases. However, this agency reported that a high percentage of countries in the region have not defined metrics for measuring progress for SDG targets 6.6, 6.a and 6.b, while available data from many countries are unreliable or incomplete, which casts doubts on the possibility of effectively monitoring progress towards meeting the SDGs in the region (ECLAC, 2017).

In this chapter describing the prospects for meeting the SDG 6 objectives in the Paraná River Basin, it was not possible to obtain data that are specific to the basin. Therefore, the discussion will focus on national data that are available for Brazil, Argentina and Paraguay, complemented with additional data from national sources and recent research. While these parameters may vary across different regions within the three countries, the data on a national level can give a general idea of the prospects within the basin for meeting the SDG 6 targets. However, we have also used sub-national data to highlight some relevant issues and support the arguments.

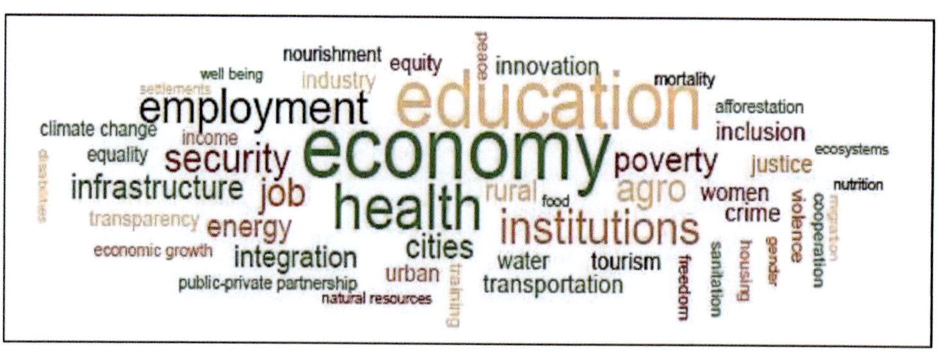

Figure 9.2 Lexicometric analysis of key concepts present in national development plans for countries in Latin America.

Source: ECLAC (2018).

The SDG process in countries of the Paraná River Basin

In May 2016, at the thirty-sixth session of ECLAC, held in Mexico City, the Commission's Member States established the Forum of the Countries of Latin America and the Caribbean on Sustainable Development as a regional mechanism to monitor and review the implementation of the 2030 Agenda for Sustainable Development. The Forum is convened under the auspices of ECLAC and meets annually to review progress at the national and regional level. In the three annual reports published to date by ECLAC, aggregate metadata were provided that track early progress in achieving the SDG goals (ECLAC, 2017, 2018 , 2019). Included in these reports are descriptions of the institutional approaches taken by the Member States to coordinate national responses to the 2030 Agenda. The countries of Brazil, Argentina and Paraguay have all established high-level institutions that are charged with coordinating the national responses. Below are the names of these institutions (translated into English) and a description of their composition and mandates:

- Brazil established the *National Commission for the Sustainable Development Goals*, which reports to the Government Secretariat of the Office of the President and includes representatives from the Office of the President, the Ministry of Foreign Affairs, the Ministry of Social and Agrarian Development, the Ministry of Planning, Development and Management and the Ministry of the Environment. Also included as technical advisors are representatives from the Geographical and Statistical Institute and the Institute of Applied Economic Research in Brazil. For the first term of the National Commission, 16 representatives from the Federal Government of Brazil, state and municipal governments and civil society were selected to participate in its activities (ECLAC, 2018).
- In Argentina, the coordinating role for the 2030 agenda was transferred to the *National Council for Social Policy Coordination*, which was originally established in 2002 and is attached to the Office of the President. This Council coordinates the actions of 20 ministries through 6 commissions, which include: education, science and technology, sustainable agricultural production, housing, urban development and infrastructure, work and employment and social protection (ECLAC, 2018).
- Paraguay established a new institution, the *Inter-Agency Coordinating Committee for the Implementation, Follow-up and Monitoring of the International Commitments Accepted by the Country in the Framework of the United Nations Sustainable Development Goals*. The Committee members include the Technical Secretariat for Economic and Social Development Planning, the Ministry of Trade, the Ministry of Foreign Affairs (coordinator) and the Social Cabinet of the President. The Committee is in the process of empowering 17 national institutions to take responsibility for the Sustainable Development Goals (ECLAC, 2017).

It is worth noting that the composition of these high-level institutions is heavily weighted towards economic development, trade, large-scale agriculture, forestry and mining, education, urban and rural planning and foreign affairs. Therefore, it is likely that the SDGs that focus on these areas of development will take precedence in the planning process, and that many of the targets for SDG 6 will be considered of lesser importance in Brazil, Argentina and Paraguay. Also, significant political changes, particularly in Brazil and Argentina, have introduced far-reaching reforms that may negatively affect the countries' ability to meet the SDGs. For instance, in Argentina, during the period 2015–2019, the national government radically abandoned the target of reducing social inequality. A similar path was taken in 2019 by the new Brazilian government.

As ECLAC's reports (2017, 2018, 2019) have clearly identified, structural inequalities have historically been a major obstacle for development in the region, and the recent reversal by the governments of the region from their commitments to reduce these inequalities may impede the achievement of the SDGs. ECLAC identified a major challenge for the Member States, which is to establish and then generate data on the SDG indicators at the national level. The current levels of production of the 232 SDG indicators by Brazil, Argentina and Paraguay are 31%, 28% and 20%, respectively (ECLAC, 2018). Several countries reporting to ECLAC have developed on-line platforms for SDG tracking, including Argentina (www.odsargentina.gob.ar/) and Brazil (www.agenda2030.com.br/), but Paraguay had not done so by the time of publication of the third report from this agency (ECLAC, 2019).

Status of SDG 6 indicators in countries of the Paraná River Basin

Meeting the targets for clean water (SDG 6.1) and sanitation (SDG 6.2)

The SDG Global Database of the UN Statistics Division (UNSD, 2019) contains global information that has been gathered over the period between approximately 2000 and 2015 related to the targets for the SDGs. The origin of these data is the WHO/UNICEF Joint Monitoring Programme for Water Supply, Sanitation and Hygiene (JMP), which along with the UN-Water Global Analysis and Assessment of Sanitation and Drinking Water (GLAAS) and the UN-Water Organization for Integrated Monitoring of Water (GEMI) are the agencies mandated to compile country data on the SDG 6 global indicators. Together these agencies make up the UN-Water Integrated Monitoring Initiative for SDG 6 (IMI). Table 9.1 shows the data gathered for indicators related to the targets for SDG 6.1 and 6.2 in each country for the most recent year in which data were collected, which is 2015.

Table 9.1 Indicator data related to the SDG 6.1 and 6.2 targets in Brazil, Argentina and Paraguay.

Targets and Related Indicator Data	*Argentina*	*Brazil*	*Paraguay*
Indicator data related to Target 6.1 By 2030, achieve universal and equitable access to safe and affordable drinking water for all			
Proportion of population using safely managed drinking water services (%); Urban areas only.	**98.5 (2015)**	**97.5 (2015)**	**NA**
Indicator data related to Target 6.2 By 2030, achieve access to adequate and equitable sanitation and hygiene for all and end open defecation, paying special attention to the needs of women and girls and those in vulnerable situations			
Proportion of population practicing open defecation (%); Urban areas only.	**1.5 (2015)**	**2.0 (2015)**	**0.4 (2015)**
Proportion of population using safely managed sanitation services (%); Urban areas only.	**26.5 (2015)**	**40.4 (2015)**	**NA**
Proportion of population with basic handwashing facilities on premises (%); All areas.	**NA**	**NA**	**NA**

Data source: UNSD (2019). The most recent date when the data were collected is shown in brackets. NA = Data not available.

The official data provided to UNSD that are presented in Table 9.1 indicates that Argentina and Brazil have essentially met a target for 100% of the population to have access to '*safely managed*' drinking water. However, there are important caveats. The data presented in the table is only for urban areas, but the situation in rural areas presents a significant challenge for both countries. Also, the official data provided by the countries must be carefully scrutinized, as discussed below.

- Argentina. The data base indicates that in Argentina the level of full access to safely managed drinking water would be essentially the same for urban areas and '*all areas*' (i.e. rural and urban). However, the available evidence gives a very different picture. For example, a recent study identified 2,432 informal urban settlements across the country that house around 3 million people living in conditions of poverty, extreme poverty and vulnerability (Argentina Ombudsman, 2018). This includes 42 settlements within the capital city of Buenos Aires. The study found that only 5% of the settlements has a formal connection to a water supply network, and the majority relies on irregular connections and other alternatives (Argentina Ombudsman, 2018). It can hardly be argued that the water available for these settlements is 'safely managed', and the evidence suggests that there is scant government control over this situation and a lack of reliable information. One of the worst affected areas is the Great North Region, housing around 20% of the country's population and covering nine provinces; many within the Paraná River Basin. This region has been historically affected by poverty and lack of investment in basic infrastructure, where official levels of coverage for water supply may be as low as 75%, as in the provinces of Misiones, Chaco or Formosa, which are located in the Paraná River Basin. Beyond the data on coverage officially reported, the services in the Great North Region (NGA) tend to be severely affected by poor quality, intermittent delivery and other problems (Argentina, Ministry of Federal Planning, Public Investment, and Services, 2012), which include the threat to human health posed by the consumption of unsafe water contaminated with naturally occurring arsenic in large areas of the NGA. Arsenic of geological origin is also a major problem affecting the safety of water supplies in large areas of the country, including provinces located in the Paraná River Basin, such as Santa Fe, where a significant proportion of the population receives water with arsenic levels that greatly exceed international standards of water safety (AWR, 2018). Also, Indigenous communities in Argentina are particularly affected by the lack of safe water supply. A study of 32 Indigenous communities scattered across ten provinces found that only 28% had access to a basic water supply (Argentina Ministry of Federal Planning, Public Investment, and Services, 2011). Unfortunately, the policies implemented in the last few years have not been focused on reversing the conditions of extreme poverty and inequality underpinning this situation and it is likely that in some areas the conditions may have worsened rather than improved.
- Brazil. There are no data in the JMP data base for rural populations in Brazil, but the rural population of the country, which is 30 million people according to the 2010 national census (about 16% of the total population) is the worst affected by the lack of safely managed drinking water. For example, *quilombola* communities of Afro-American descent, whose actual number is unknown but placed by official estimates as at least at 600,000 distributed in around 2,300 rural communities scattered across several states, have scant access to safely managed drinking water. A recent study that looked at 173 *quilombola* communities reported that only 6.4% had access to 'treated drinking water'

(de Pádua et al., 2015). In urban areas the situation is also problematic, because even if the water supplied by the utilities were 'safely managed', the evidence shows that too often the water quality is compromised by ageing and deteriorated infrastructure to the point that when it reaches the households it is no longer safe for human consumption. As pointed out by Marcos Montenegro, President of the Brazilian Association of Sanitary Engineering in Brasilia, there is a lack of credible information in the country about the quality of water distributed to the population (Montenegro, 2015). Recent developments suggest that the situation may have worsened, as the policies implemented by the national government since 2015 have weakened the public institutions in charge of planning and monitoring the delivery of essential water and sanitation services (ONDAS, 2019).

- Paraguay. This country has been praised for making rapid progress in extending coverage for water supply in recent decades. However, JMP does not have any data for the country's provision of safely managed drinking water (Table 9.1). A recent WHO-UNICEF joint report presented official government data for 2015, stating that 89% of the national population had access to piped water, a figure that reached 95% in urban areas but dropped to 81% in rural areas. This report also lacked information on the proportion of drinking water supplied that was 'safely managed' or 'free from contamination' (WHO-UNICEF, 2017). Overall, the reliability and consistency of reported official figures are a matter for concern. A recent government report estimated that in 2016 the national coverage for piped water supply was 78% but cautioned that this information was out of date and that the real figure was likely lower. The report also stated that only 79.1% of the population receives 'disinfected water' and that an unspecified but 'important share of the Paraguayan population' drinks water 'without any guarantee of its quality' (Paraguay Ministry of Public Works and Communications, 2018). The worst affected are marginalized urban areas and particularly rural areas, which house about 40% of the country's population, including Indigenous communities.

Among the countries within the Paraná River Basin, an important aspect on reaching the SDG 6 goals for access to safe drinking water are the inequalities in access between different groups. ECLAC (2018) reported data on the gaps in drinking water coverage between the highest and lowest income households in 17 Latin American countries and these data show gaps for urban populations of approximately 7%, 3% and 1% for Paraguay, Brazil and Argentina, respectively, and for rural populations, gaps of approximately 10% and 25% in Paraguay and Brazil, respectively (no data for Argentina). Clearly, the challenges of providing access to safe drinking water are most acute among the rural poor.

In relation to the target for 100% access to facilities for sanitation and hygiene (i.e. SDG target 6.2), there are data in the JMP data base for three indicators that are related to this target (Table 9.1). According to these data, all three countries in the basin have essentially eliminated open defecation in urban areas, although in 2017 Brazil still reported 11% open defecation in rural areas (UNSD, 2019). There will be significant challenges for the three countries to meet the targets for 100% access to sanitation services, although there are no data available in the data base to ascertain the actual situation in Paraguay. Yet, an official government report recently reported that outside the capital city of Asunción and the Central District where there is 'greater coverage', in all other regions of the country the sanitation coverage is at best 10% and many places lack any sanitation facilities (Paraguay Ministry of Public Works and

Communications, 2018). In addition, JMP has not recorded data for any of the three countries on the proportion of the population with handwashing facilities on the premises. ECLAC (2018) reported that the gaps in sanitation coverage between the highest income and lowest income households in rural areas in Paraguay and Brazil were approximately 65% and 27%, respectively (no data for Argentina), so, similar to the situation noted in the ECLAC report on access to safe drinking water, the greatest challenge for providing access to sanitation services will be among the rural poor. However, access to sanitation services is also a challenge in urban areas among the lowest-income segment of the population. In Brazil, Argentina and Paraguay, between 19–21% of urban populations live in slums where there is poor access to sanitation services (ECLAC, 2018).

Summing up, Argentina and Brazil officially reported to the JMP very high levels of coverage for safe drinking water, but a critical scrutiny of the data shows that the information is incomplete and does not reflect the realities on the ground. The situation is probably similar in Paraguay, in relation to the reliability of the available information. With lower levels of reported coverage for safe drinking water, Paraguay is likely to experience greater difficulties to achieve the universalization of safely managed drinking water by 2030. The three countries will struggle to meet the target for sanitation, and Brazil will have a particularly daunting challenge if it is to reduce the high proportion of people practicing open defecation in rural areas, where 16% of its population is located. In this regard, for all three countries that share the Paraná River Basin, the greatest challenges to meet the targets of SDG 6.1 and 6.2 will be to provide these services to the segments of the population that make up the rural poor and the urban poor. Yet, there are significant obstacles that cast serious doubt on the ability of these countries to meet the 2030 targets. In particular, the introduction since 2015 of regressive policies in the organization and delivery of essential public services in Argentina and Brazil provides reason for concern, given that meeting the 2030 targets will require a strong State commitment to reduce structural inequalities in order to provide access to essential services to the poorest sectors of the population. Unfortunately, the evidence suggests that the policies in place in both countries are widening the inequality gap and very likely will make it impossible to meet the SDG 6 targets for water, sanitation and hygiene. The situation in Paraguay is less clear, owing to the lack of precise information, but there are few reasons for optimism.

Having said that, a possible failure to meet the SDG 6 targets for water, sanitation and hygiene does not mean that there will not be some progress, even if it is slower and less extensive than expected. The data in Figure 9.3 show the increase in access to 'safely managed' sanitation services and the decline in open defecation that occurred between 2000 and 2015 among rural and urban populations in Brazil, as reported to the JMP data base. These data show the slow but measurable progress in achieving these targets over the 15-year monitoring period. However, the trends also illustrate the gap between urban and rural populations. Only the data for Brazil are provided here because Argentina and Paraguay did not report data for its rural populations and there are no data on access to sanitation services for Paraguay.

Meeting other SDG 6 targets

Table 9.2 summarizes the most recent indicator data available in the UN data base related to the targets for SDG 6.3, 6.4, 6.5 and 6.6. It is the responsibility of the GEMI program to gather these data under the umbrella of IMI. The Executing Agency for the GEMI program is

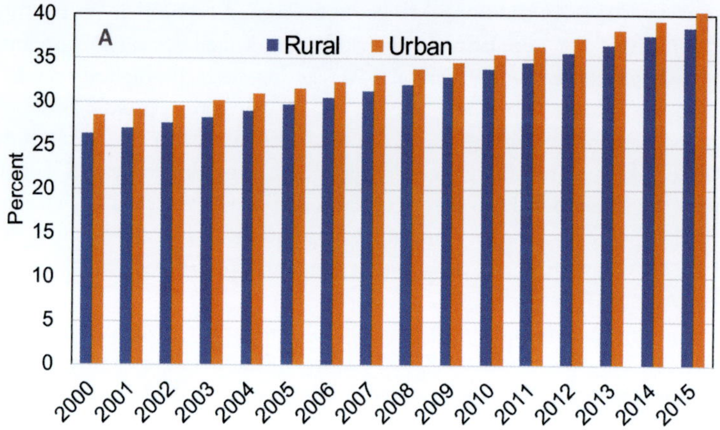

A) Sanitation

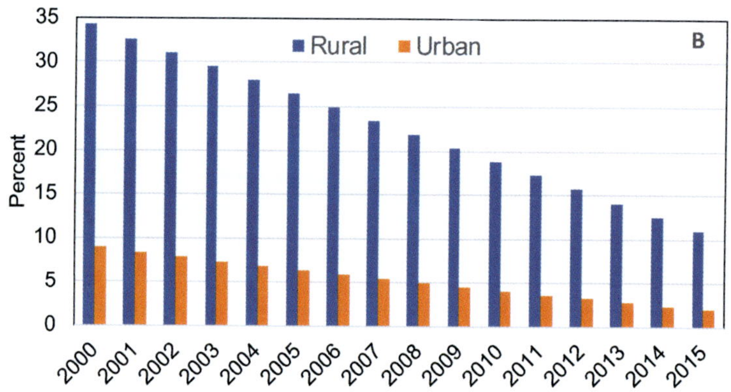

B) Defecation

Figure 9.3 Temporal trends (2000–2015) in Brazil for: (A) percentage of the rural and urban populations with access to safely managed sanitation services (%) and (B) the percentage of the rural and urban populations practicing open defecation.

Data source: UNSD (2019).

GEMS/Water, which includes three operating centers, the Global Programme Co-ordination Unit, the Capacity Development Centre and the GEMStat Data Centre. SDG 6.3 addresses the critical element of water quality protection and is closely interconnected with SDG 6.1 and 6.2. To monitor progress in achieving this target, two key indicators have been identified. Countries that set national aspirations for Indicator 6.3.1 for treatment of wastewater must monitor their progress in increasing the proportion of wastewater that is 'safely treated'. Under Indicator 6.3.2, countries must define their targets for increasing the proportion of 'bodies of water with good ambient water quality', where 'ambient' refers to water quality in rivers, lakes and ground water, and water of 'good' quality is judged to be sufficient to maintain ecosystem functions and not be a risk to human health. Each country can set their own water quality guidelines for assessing whether the quality of their surface waters can be described as 'good'. Eventually, global water quality standards may be developed as benchmarks for these national standards. Table 9.2 summarizes the current data in the GEMI data base related to the indicators for SDG 6.3 in Brazil, Argentina and Paraguay.

Table 9.2 Indicator data related to the targets for SDG 6.3, 6.4, 6.5 6.6, 6.a and 6.b in Brazil, Argentina and Paraguay.

Targets and indicators	*Argentina*	*Brazil*	*Paraguay*
Target 6.3 By 2030, improve water quality by reducing pollution, eliminating dumping and minimizing release of hazardous chemicals and materials, halving the proportion of untreated wastewater and substantially increasing recycling and safe reuse globally			
Proportion of wastewater safely treated (%)	**22.5 (2018)**	**33.5 (2018)**	**NA**
Proportion of bodies of water with good ambient water quality (%)	**NA**	**73% (2017)**	**NA**
Target 6.4 By 2030, substantially increase water-use efficiency across all sectors and ensure sustainable withdrawals and supply of freshwater to address water scarcity and substantially reduce the number of people suffering from water scarcity			
Change in water-use efficiency over time (USD/m^3)	**12.1 (2010)**	**NA**	**6.9 (2010)**
Level of water stress: freshwater withdrawal as a proportion of available freshwater resources (%)	**10.5 (2010)**	**3.0 (2015)**	**1.8 (2010)**
Target 6.5 By 2030, implement integrated water resources management at all levels, including through transboundary cooperation as appropriate			
Degree of integrated water resources management implementation (0–100)	**38 (2017)**	**51 (2017)**	**32 (2017)**
Proportion of transboundary basin area (lakes, rivers, aquifers) with operational arrangement for cooperation (%)	**NA**	**62.4 (2017)**	**50.9 (2017)**
Target 6.6 By 2020, protect and restore water-related ecosystems, including mountains, forests, wetlands, rivers, aquifers and lakes			
Change in the extent of water-related ecosystems over time (loss, % of the baseline value)	**NA**	**NA**	**NA**
Target 6.a By 2030, expand international cooperation and capacity-building support to developing countries in water and sanitation-related activities and programmes, including water harvesting, desalination, water efficiency, wastewater treatment, recycling and reuse technologies			
Total official development assistance (gross disbursement) for water supply and sanitation, by recipient countries (millions of constant 2016 USD)	**5.5 (2016)**	**105.0 (2016)**	**18.1 (2016)**
Target 6.b Support and strengthen the participation of local communities in improving water and sanitation management			
Countries with procedures in law or policy for participation by service users/communities in planning program in rural drinking-water supply.	**Clearly defined (2017)**	**Clearly defined (2017)**	**Clearly defined (2017)**

(Continued)

Table 9.2 (Continued)

Targets and indicators	*Argentina*	*Brazil*	*Paraguay*
Sub-indicator: Countries with users/ communities participating in planning programs in rural drinking-water supply, by level of participation.	**Low (2017)**	**Moderate (2017)**	**Moderate (2017)**
Sub-indicator: Countries with procedures in law or policy for participation by service users/communities in planning programs in water resources planning and management, by level of definition in procedures.	**Clearly defined (2017)**	**Clearly defined (2017)**	**Clearly defined (2017)**
Sub-indicator: Countries with users/ communities participating in planning programs in water resources planning and management, by level of participation.	**Moderate (2017)**	**Moderate (2017)**	**Moderate (2017)**

Data source: UNSD (2019). The most recent date when the data were collected is shown in brackets. NA = Data not available.

Effective treatment of municipal and industrial wastewater is a key driver for improving water quality, especially in countries such as Brazil, Argentina and Paraguay, where rapid urbanization, population growth and industrialization have increased the need for treatment of wastewater. As shown in Table 9.2, the proportion of wastewater 'safely treated' in Argentina and Brazil is currently below 35%. There are no data available in the UNSD data base for Paraguay, although a recent government report states that wastewater treatment is available only in the capital and a few other cities, and that it is limited to primary or at most secondary treatment (Paraguay Ministry of Public Works and Communications, 2018). Wastewater treatment in these countries is clearly below the level needed to protect the quality of aquatic resources. Table 9.3 shows the proportions of safely treated wastewater that exceed a level of 35% in selected other countries around the globe. If Brazil, Argentina and Paraguay are to increase the proportion of effective wastewater treatment to even the current level in Chile (72.5%) and Mexico (51%), significant investment is needed to install new systems or to upgrade existing infrastructure for the treatment of municipal wastewater.

Table 9.3 Selected countries where the proportion of safely treated wastewater (%) exceeds 35%.

Country	*Safely Treated Wastewater (%)*
Switzerland	99
Germany	98
USA	90
Italy	88
Canada	79.5
Australia	73
Chile	72.5
Mexico	51
China	45
Peru	39

Data source: UNSD (2019).

As stated by Fernandez Cirelli and Ojeda (2008) in describing wastewater management in Argentina, 'the existence of a sewerage system does not mean automatically wastewater treatment'. In a review of the reliability of wastewater treatment systems in Brazil, the authors concluded that the traditional process of designing and operating wastewater treatment systems focuses on the technology, and too little emphasis is placed on operator training and providing the resources needed to maintain the facilities (Oliveira and Von Sperling, 2008). In Paraguay, most of the existing wastewater treatment systems are not functioning properly (Cuppens et al., 2013). Some argue that centralized wastewater treatment systems that require an extensive sewerage network and technologically complex treatment systems are not appropriate in all situations; especially in lower- and middle-income countries. Sewage collection systems account for over 60% of the budget for wastewater management in a centralized system (Massoud et al., 2009). Decentralized or 'cluster' systems where wastewater is collected for treatment from a small number of households may be a more appropriate and affordable treatment option in many communities (Nansubuga et al., 2016).

As shown in Table 9.2, data on water quality are only available in the UNSD database for Brazil, and these data indicate that 73% of water bodies in that country are classified as having good ambient water quality. The targets for water quality under SDG 6.3 are aspirational, as each country can set their own criteria for ambient water quality. However, the UN-Water guidelines recommend six 'core' parameters as a basic data set for monitoring water quality in rivers, lakes and in groundwater (Table 9.4). In addition, other '*progressive monitoring*' parameters are recommended for data collection, if warranted and/or the country has the capacity to measure them. The overall strategy of UN-Water is to encourage countries to expand their monitoring programs, increase the range of water quality parameters and adopt more sophisticated approaches for calculating water quality indexes. The level of spatial and temporal resolution of the monitoring program is also defined by the countries at a national level. The number of sampling stations and the frequency of monitoring are flexible and can be increased as the country develops capacity and resources. However, the UN-Water methodology recommends that monitoring should be at least four times per year at each station. Countries have the option of maintaining their own data base for water quality parameters and reporting their results to GEMI, or they can submit their water quality data to the GEMStat central data base. The global monitoring locations illustrated in Figure 9.4 indicate that Brazil is providing extensive water quality information through the GEMStat central data base. There are data from a few monitoring stations in Argentina that have been reported to GEMI, but no water quality data have been submitted to GEMI from Paraguay (Figure 9.4). The available information on

Table 9.4 The core parameters identified for water quality monitoring in rivers, lakes and groundwater (GW) using the SDG Indicator 6.3.2 methodology. The 'X' notation indicates the core parameters to be monitored within rivers, lakes and groundwater.

Core Parameter	*River*	*Lake*	*GW*
Dissolved oxygen	×	×	
Electrical conductivity			×
Total oxidized nitrogen	×	×	
Nitrate			×
Orthophosphate	×	×	
pH	×	×	×

Source: UN-Water (2017).

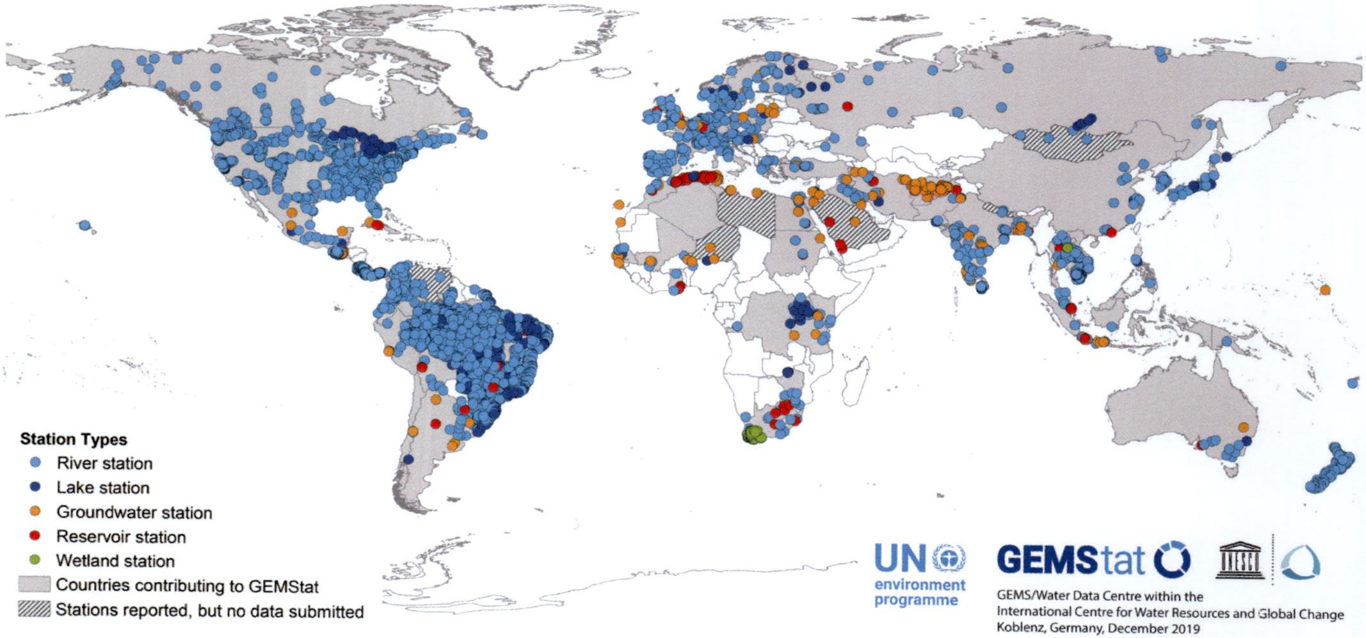

Figure 9.4 Global distribution of the water quality monitoring stations included in the GEMStat database of GEMS/Water (2017). The figure was provided by the International Centre for Water Resources and Global Change, Koblenz. Available upon request from GEMS/Water Data Centre: data-request@gemstat.org.

water quality within the Paraná River Basin is reviewed in Chapter 7 in this book, and the key message from this chapter is that anthropogenic activities (e.g. agriculture, urbanization, hydroelectric development) are impairing the quality of water in the Paraná River Basin.

The UNSD data base includes reports from several countries on their water quality. The proportions of water bodies classified as having good water quality in Germany (72%), South Africa (62.5%), the Netherlands (53%) and Sweden (49%) are all lower than the proportions reported by Brazil (i.e. 73%). Since the definition of 'good' water quality and the scope and frequency of the monitoring programs are set by each country, it is difficult to compare water quality data across countries. However, clearly Brazil has recognized the importance of developing a comprehensive water quality monitoring program in order to assess progress in meeting the targets for SDG 6.3. In Argentina, the Ministry of the Interior maintains an on-line data base for a network of hydrological stations within the country, and there are some water quality parameters reported in the data base for selected sites in the Paraná basin, mainly for years since 2010: http://bdhi.hidricosargentina.gob.ar. Paraguay needs to develop monitoring programs in order to fully participate in the SDG process. The qualitative data presented in Table 9.2 on data related to Targets 6.a and 6.b indicate that all three countries that share the Paraná River basin have the institutional, regulatory and organizational tools to make progress in improving water quality. However, as emphasized in Chapter 2 in this book on the subject of water management in the basin, political will is required to make the necessary investments.

According to the data related to attempts to increase water use efficiency (Table 9.2), increasing the efficiency of water use is not a high priority in Brazil and Paraguay, probably because at a country level there are abundant surface water and groundwater resources. According to the data for the level of water stress, as indicated by freshwater withdrawals as a proportion of available freshwater resources, water withdrawals in Argentina are currently at a level of 10.5% of available resources (Table 9.2). However, it is probable that the highest rate of withdrawals relative to resources is in the more arid southern and western provinces of the country, rather than in the northeastern areas of Argentina within the Paraná River Basin. The reported level of water stress is less than 5% in Paraguay and Brazil (Table 9.2). In the latest UN report on the status of the SDGs, 22 countries reported levels of water stress above 70%, which is a strong indicator of future water scarcity in these regions (United Nations, 2018). Compared to these countries, according to the data from the UN report, the nations that share the Paraná River Basin would not be under a high level of threat from water scarcity. The surface water and groundwater water resources in the Paraná River Basin are described in detail in Chapter 2 in this book.

However, the national data on the levels of water stress obscure the very high inter-regional variability of available water resources, and the fact that large areas of these countries, particularly in Argentina and Brazil, are subject to water stress.

- Argentina. Two thirds of the territory of Argentina are arid or semi-arid, are subject to high inter-annual rainfall variability and in some cases, record severe levels of water stress (Bianchi and Cravero, 2010). This is compounded by the fact that the scarce water resources in these regions are under threat from heavy pollution caused by large-scale mining and other industrial activities, while groundwaters in many areas are affected by naturally high levels of arsenic, boron and other substances that compromise the water quality for human uses (Calcagno and Mendiburo, 2000; AWR, 2018). Moreover, the impact of deforestation and global warming have been identified as the main factors behind recurrent extreme drought events recorded in the last two decades, which are expected to become a regular feature (ONDTyD, 2019).

- Brazil. Despite having the world's greatest availability of freshwater, Brazil has very high variability in the distribution of water resources. Historically, the country's large northeastern region has been characterized by extreme aridity and regular, and even permanent drought conditions. However, owing to processes of deforestation, desertification and urbanization, the areas at risk of water stress have been expanding and now include the country's largest metropolises, São Paulo, Rio de Janeiro and Brasilia among others, which have been classified as being at 'high' and 'extreme risk' of water stress, on the same level as some Middle Eastern countries (WRI, 2019). The country has a monitoring system for droughts that provides up to date information and that is continuously expanding the regions of the country being monitored for drought occurrences (ANA, 2019; CEMADEN, 2019).
- Paraguay. Although the country is also classified as at low risk for water stress owing to its water abundance, 34% of its territory is arid. It is also characterized by regional diversity and is subject to similar pressures from deforestation, desertification and pollution of water sources. In particular, the region known as Gran Chaco, which is shared by Argentina, Bolivia, Brazil and Paraguay (20% of this region is in Paraguayan territory), is characterized by aridity and extreme aridity. This region is also home to several Indigenous communities, which have difficulty in accessing safe and sufficient water supplies. The region is under great threat from the expansion of activities that extract high volumes of water, particularly large-scale agriculture and mining, with deforestation and pollution of water sources also being a major problem (UNESCO, 2010; PAHO, 2011).

SDG 6.5 focuses on implementation of integrated water resources management at the national level, and through transboundary cooperation. Integrated water resource management has been promoted by the UN as an approach to improve global water stewardship. In the latest UN report on the status of the SDGs, 157 countries reported an average implementation of integrated water resources management of 48% and based on data from 62 countries sharing transboundary waters, the average percentage of national transboundary basins covered by an operational arrangement was only 59% (United Nations, 2018). The data summarized in Table 9.2 show that the countries that share the Paraná River Basin are performing at about the average or slightly below the average among the nations that reported statistics to the global data base. The current approaches to transboundary cooperation in the basin are described in Chapter 8 in this book, but most of these are bilateral agreements that focus on access to hydroelectric power and not to the protection of water resources.

The Organization of American States (OAS), in partnership with the ANA in Brazil, the Global Environmental Facility of the World Bank and UNEP is currently supporting a project entitled, Implementation of Integrated Watershed Management Practices for the Pantanal and Upper Paraguay River basin, which aims to protect habitat in this region: www.oas.org/dsd/ALTOPARA/Pd.htm#Expected%20Results.

However, there are several barriers to implementing integrated water resource management, and past applications of this approach in the real world have 'left much to be desired' (Biswas, 2008). One of the problems is the lack of incentives among stakeholders to implement these measures. As an example, an analysis of a proposed payment for ecosystem services scheme for the Pantanal wetland in Brazil concluded that 'large-scale implementation is unlikely' because of socioeconomic inequality between the inhabitants of the region and a lack of incentives for wealthy farmers in the uplands (Schulz et al., 2015). Therefore, new approaches may be needed to promote the protection of aquatic resources. Some argue that integrated water resource management should be abandoned for more pragmatic approaches that include financial models, such as the 'water-energy-food-climate nexus' (Schmidt and Matthews, 2018). In addition, we cannot ignore the potential impact of climate change on water resources within the basin, which relates to SDG 13. Based on predictions from climate change models, the authors of Chapter 5 in this book concluded that the northern part of the basin will become drier, with reduced flows in the Upper Paraná.

The focus of SDG 6.6 is to protect and restore water-related ecosystems, including mountains, forests, wetlands, rivers, aquifers and lakes. Chapter 4 in this book describes the water-related ecosystems within the basin, including the extensive wetland ecosystems that are threatened by agricultural development and water management schemes. The three countries that share the Paraná River Basin collectively have huge areas of permanent water bodies to protect. However, no data have been submitted to GEMI by Brazil, Argentina or Paraguay on progress towards restoring water-related ecosystems. Moreover, we have now reached the time horizon for meeting this SDG indicator, as 2020 is the end date for this target. Because of the synergies between the Aichi Biodiversity Targets and the SDG target 6.6, this earlier deadline was selected to coincide with the end date of the Strategic Plan on Biodiversity (2011–2020). Both Paraguay and Argentina are parties to the Aichi agreement, but according to the on-line data base that tracks national objectives under the agreement, these countries have not set targets for protection of biodiversity under the program (Convention on Biological Diversity, 2019). However, Brazil has been active in setting targets for this process. For instance, under Article 5, Brazil has pledged to reduce by at least 50% (in comparison with the 2009 rate) the rate of loss of native habitats, and to significantly reduce degradation and fragmentation in all biomes. Whether progress will have been made towards these goals will be a subject for discussion at the UN Biodiversity Conference to be held in October 2020 in Kunming, China. However, expectations are that most of the Aichi targets set by participating nations will not be achieved by the time of the conference (Tsioumani, 2019).

Conclusions

According to the information officially reported to JMP, the nations that share the Paraná River basin have essentially met the basic targets of SDG 6 to provide universal access to safe drinking water and to sanitation services. However, on closer scrutiny of the available information, it becomes evident that there are many reasons to call into question the actual level of achievement and the likelihood that these countries will meet the targets by 2030. First, the coverage of 'safely managed' drinking water is open to question, as the evidence suggests that it is much lower than the levels officially reported by Argentina and Brazil. Despite the lack of reliable data from Paraguay, available official information recognizes that the country is also lagging behind in this area. In all three cases, the main problems are concentrated in poor urban sectors and in rural areas, and particularly affect Indigenous communities and Afro-American communities in Brazil. There will be very significant challenges for these countries to meet the SDG 6 targets of treating municipal wastewater and improving water quality.

The rapid reduction of extreme poverty achieved in Argentina and Brazil during the period 2008–2015 had raised expectations that economic growth and further poverty reduction would reduce the gap between rich and poor in accessing basic services. However, radical political changes that have taken place since 2015 in the region have triggered a retreat from policy interventions oriented at tackling inequalities and improving living conditions through redistributive policies and strong investments in infrastructures for essential public services. Instead, the situation in both countries, and notably in Argentina, shows an enormous regression with a steep rise in poverty levels by close to 40% by November 2019 and a rapidly worsening inequality gap. These conditions will certainly have an impact on the abilities of countries within the basin to deliver on meeting the targets of SDG 6, since these will require leadership and investment at federal, regional and local levels in Brazil, Argentina and Paraguay.

An additional concern, as indicated by the organizational structures of the national institutions that have been charged with acting on the 2030 Agenda in the three countries, is that making progress in SDG 6 may not be a high priority relative to other development goals, particularly those related to economic growth. Still, the SDG targets for reducing poverty, improving education and promoting health and wellbeing are compatible and mutually reinforcing with at least the basic components of SDG 6, access to safely managed drinking water, sanitation, wastewater treatment and hygiene. From another perspective, and very important for the monitoring process, the annual meetings of the Forum of the Countries of Latin America and the Caribbean on Sustainable Development may provide impetus for these countries to gather and share their data on progress in meeting the SDG 6 targets.

References

ANA. 2019. *Drought Monitor* (in Spanish). National Water Agency, Brasilia, Brazil. Retrieved from: http://monitordesecas.ana.gov.br.

Argentina Ministry of Federal Planning, Public Investment, and Services. 2011. *Project Water Infrastructure for the Great North II. Planning Framework for Indigenous Communities* (in Spanish). Buenos Aires, Argentina. Retrieved from: www.ucpypfe.gov.ar/BirfPIHNG/MPPI-2.pdf.

Argentina Ministry of Federal Planning, Public Investment, and Services. 2012. *Development Programme for the Provinces of the Great North, Project Profile* (in Spanish). Buenos Aires, Argentina. Retrieved from: http://idbdocs.iadb.org/wsdocs/getdocument.aspx?docnum=36801963.

Argentina Ombudsman (Defensoría del Pueblo de la Nación). 2018. *Informal Settlements and Human Rights, Buenos Aires, Argentina.* Retrieved from: www.ohchr.org/Documents/Issues/Housing/InformalSettlements/ArgentinaDefensorPueblo.pdf.

AWR. 2018. *Arsenic in Water, Final Report* (in Spanish). Arsenic in Water Research Group, Buenos Aires, Argentina. Retrieved from: https://rsa.conicet.gov.ar/wp-content/uploads/2018/08/Informe-Arsenico-en-agua-RSA.pdf.

Bianchi, A.R., Cravero, S.A.C. 2010. *Atlas Climático Digital de la República Argentina.* National Institute of Agricultural Technology (INTA), Buenos Aires. Retrieved from: https://inta.gob.ar/documentos/atlas-climatico-digital-de-la-republica-argentina; Accessed on March 8, 2020.

Biswas, A.K. 2008. Integrated water resource management: Is it working? *Water Resources Development* 24:5–22.

Calcagno, A., Mendiburo, N., 2000 *Report on water management in the Argentine Republic* (in Spanish). Economic Commission for Latin America and the Caribbean (ECLAC), Santiago de Chile. Retrieved from: www.cepal.org/drni/proyectos/samtac/inar00200.pdf.

CEMADEN - National Monitoring and Alerting Centre for Natural Disasters. 2019. *Monitoring drought in Brazil.* Retrieved from: www.cemaden.gov.br/categoria/monitoramento/seca-no-brasil/.

Convention on Biological Diversity. 2019. Retrieved from: www.cbd.int/sp/targets/.

Cuppens, A., Smets, I., Wyseure, G. 2013. Identifying sustainable rehabilitation strategies for urban wastewater systems: A retrospective and interdisciplinary approach. Case study of Coronel Oviedo, Paraguay. *Journal of Environmental Management* 114:423–432.

de Pádua, V.L., Vilela, D.R., Silva, A.S.R. 2015. Participative generation of a water treatment system in a Quilombola community in Minas Gerais, Brazil. *WATERLAT-GOBACIT Working Papers*, 2(12). Retrieved from: http://waterlat.org/WPapers/WPSPIDES212.pdf.

ECLAC. 2017. *Annual report on regional progress and challenges in relation to the 2030 Agenda for Sustainable Development in Latin America and the Caribbean, Economic Commission for Latin America and the Caribbean (ECLAC)*, Forum of the Countries of Latin America and the Caribbean on Sustainable Development, Mexico City, Mexico, April 26–28, 2017, 111 p. Retrieved from: https://foroalc2030.cepal.org/2017/en.

ECLAC. 2018. *Second annual report on regional progress and challenges in relation to the 2030 Agenda for Sustainable Development in Latin America and the Caribbean, Economic Commission for Latin America and the Caribbean (ECLAC)*, Forum of the Countries of Latin America and the Caribbean on Sustainable Development, Santiago, Chile, April 16–20, 2018, 167 p. Retrieved from: https://repositorio.cepal.org/handle/11362/43439?locale-attribute=en.

ECLAC. 2019. *Quadrennial report on regional progress and challenges in relation to the 2030 Agenda for Sustainable Development in Latin America and the Caribbean, Economic Commission for Latin America and the Caribbean (ECLAC)*, Forum of the Countries of Latin America and the Caribbean on Sustainable Development, Santiago, Chile, April 24–26, 2019, 167 p. Retrieved from: http://caribbean.cepal.org/content/quadrennial-report-regional-progress-and-challenges-relation-2030-agenda-sustainable.

Ederington, J., Levinson, A., Minier, J. 2005. Footloose and pollution-free. *Reviews in Economic Status* 87:92–99.

Fernandez Cirelli, A., Ojeda, C. 2008. Wastewater management in Greater Buenos Aires, Argentina. *Desalination* 218:52–61.

GBD. 2016. SDG Collaborators. 2017. Measuring progress and projecting attainment on the basis of past trends of the health-related Sustainable Development Goals in 188 countries: An analysis from the Global Burden of Disease Study 2016. *Global Health Metrics* 390:1423–1459.

GLAAS. 2017. Financing universal water, sanitation and hygiene under the Sustainable Development Goals, UN-Water Global Analysis and Assessment of Sanitation and Drinking Water. Retrieved from: http://apps.who.int/iris/bitstream/10665/254999/1/9789241512190-eng.pdf?ua=1.

Massoud, M.A., Tarhini, A., Nasr, J.A. 2009. Decentralized approaches to wastewater treatment and management: Applicability in developing countries. *Journal of Environmental Management* 90:652–659.

Montenegro, M.H. 2015. Presentation by the President of the Brazilian Association of Sanitary Engineering, Federal District (ABES-DF), at Round Table 1, 'Materializing the human right to water: The challenge in perspective', A Seminar for Research and Debate – DESAFIO Project, Materializing the Right to Water and Basic Sanitation Services – Day 1, Brasilia, September 9, 2015. Retrieved from: https://youtu.be/-IiNvfgsZEk?list=PLx6qphzdSP6sxHigH5SMJSfaa31tYyl8u.

Nansubuga, I., Banadda, N., Verstrate, W., Rabaey, K. 2016. A review of sustainable sanitation systems in Africa. *Reviews on Environmental Science and Biotechnology* 15:465–478.

Oliveira, S.C., Von Sperling, M. 2008. Reliability analysis of wastewater treatment plants. *Water Research* 42:1182–1194.

ONDAS (in Portuguese). 2019. *National Observatory of the Rights to Water and Sanitation 2019*. Brasilia, Brazil. Retrieved from: https://ondasbrasil.org.

ONDTyD. 2019. *National Observatory of Land Degradation and Desertification: Buenos Aires and Mendoza, Argentina*. Retrieved from: www.desertificacion.gob.ar.

Pan American Health Organization (PAHO). 2011. *The Great South American Chaco*. Retrieved from: http://iris.paho.org/xmlui/handle/123456789/10027?locale-attribute=pt.

Paraguay Ministry of Public Works and Communications, Directorate for Potable Water and Sanitation. 2018. *National Plan of Potable Water and Sanitation*. Asunción, Paraguay. Retrieved from: www.mopc.gov.py/pnaps/pnaps.pdf.

Rocha, D.G., Weiss, V.P.A. 2019. The convergences between the Sustainable Development Goals and national agendas: The Brazilian case. *Health Promotion International* 34(S1):i46–i55.

Sapkota, P., Bastola, U. 2017. Foreign direct investment, income, and environmental pollution in developing countries: Panel data analysis of Latin America. *Energy Economics* 64:206–212.

Schmidt, J.J., Matthews, N. 2018. From state to system: Financialization and the water-energy- food-climate nexus. *Geoforum* 91:151–159.

Schulz, C., Ioris, A., Martin-Ortega, J., Glenk, K. 2015. Prospects for payments for ecosystem services in the Brazilian Pantanal: A scenario analysis. *Journal of Environment and Development* 24:26–53.

Tang, J. 2015. Testing the pollution haven effect: Does the type of FDI matter? *Environmental Resource Economics* 60:549–578.

Teodosiu, C., Robu, B., Cojocariu, C., Barjoveanu, G. (2015) Environmental impact and risk quantification based on selected water quality indicators. *Natural Hazards* 75:S89–S105.

Tsioumani, E. 2019. On the road to the 2020 UN Biodiversity Conference: Imagining the post- 2020 global biodiversity framework. SDG Knowledge Hub, May 30, 2019, International Institute for Sustainable Development, Winnipeg, Manitoba, Canada. Retrieved from: http://sdg.iisd.org/commentary/policy-briefs/on-the-road-to-the-2020-UN-biodiversity-conference/.

UNESCO. 2010. *Atlas of Arid Zones in Latin America and the Caribbean*. Montevideo, Uruguay. Retrieved from: https://unesdoc.unesco.org/ark:/48223/pf0000216333.

United Nations. 2018. *The Sustainable Development Goals Report*. Retrieved from: https://unstats.un.org/sdgs/files/report/2018/TheSustainableDevelopmentGoalsReport2018-EN.pdf.

UNSD. 2019. *Global SDG Indicators Database*. UN Statistics Division. Retrieved from: https://unstats.un.org/sdgs/indicators/database/.

UN-Water. 2016. *Water and Sanitation Interlinkages across the 2030 Agenda for Sustainable Development*. UN Water Geneva, Switzerland, 47 p. Retrieved from: www.unwater.org/publications/water-sanitation-interlinkages-across-2030-agenda-sustainable-development/.

UN-Water. 2017. Step-by-step monitoring methodology for SDG indicator 6.3.2 on ambient water quality, Chapter 4, Integrated Monitoring Guide for SDG 6, Revision January 18, 2017. Retrieved from: www.unwater.org/publications/step-step-methodology-monitoring-water-quality-6-3-2/.

UN-Water. 2018. Sustainable Development Goal 6: Synthesis Report on Water and Sanitation, Retrieved from: www.unwater.org/publication_categories/sdg-6-synthesis-report-2018-on-water-and-sanitation/.

WHO-UNICEF. 2017. *Progress on Drinking Water, Sanitation and Hygiene: 2017 Update and SDG Baselines*. World Health Organization, and United Nations Children's Fund, New York. Retrieved from: www.unicef.org/publications/index_96611.html.

Winpenny, J., Trémolet, S., Cardone, R., Kolker, J., Kingdom, B., Mountford, L. 2016. *Aid Flows to the Water Sector: Overview and Recommendations*. The World Bank, Washington, D.C..

WRI. 2019. *Ranking shows the highest risk areas in Brazil and the world for lack of water* (in Portuguese). World Resources Institute, Washington, D.C.. Retrieved from: https://wribrasil.org.br/pt/blog/2019/08/ranking-mostra-onde-ha-maior-risco-de-faltar-agua-no-brasil-e-no-mundo.

INDEX

Page numbers in **bold** refer to tables; those in *italics* refer to maps or images.